A New Deal in Dyess

Also by Van Hawkins

Hampton and Newport News
A look at two historic Virginia towns, 1975

Dorothy and the Shipbuilders of Newport News
The story of an iconic American shipyard, 1976

The Historic Triangle
How Jamestown, Williamsburg, and Yorktown
made American history, 1980

Plowing New Ground
The Southern Tenant Farmers Union
and its place in Delta History, 2007

Duty Bound
The Hyatt brothers and Confederates
of the Third Arkansas Infantry Regiment, 2011

Horizons
A novel about growing up in a small southern town
in the 1950s and 1960s, 2012

Smoke Up the River
Steamboats and the Arkansas Delta, *2016*

Moaning Low: From Slavery to Peonage
Involuntary Servitude in the Arkansas Delta, 2019

A New Deal in Dyess

*The Depression Era
Agricultural Resettlement Colony
in Arkansas*

VAN HAWKINS

A New Deal in Dyess

*The Depression Era
Agricultural Resettlement Colony
in Arkansas*

Copyright © 2020 by Van Hawkins
Revised edition
First Printing 2015

All rights reserved. No part of this publication may be reproduced, distributed, or transmitted in any form or by any means, including photocopying, recording, or other electronic or mechanical methods, without prior written permission, except in the case of brief quotations embodied in certain critical reviews and certain other noncommercial uses permitted by copyright laws.

Cover Image: Official logo for Dyess Colony, Inc.
National Archives
Records of the Work Projects Administration

Printed in the United States of America
ISBN: 978-0-9863992-3-7
Library of Congress Control Number: 2020905756

Writers Bloc
Jonesboro, Arkansas

*This book is dedicated to the Dyess folks,
who came and went,
but never forgot.*

Table of Contents

Notes and Acknowledgments	8
Prologue	13
1. A Strong Brown God's Creation	15
2. No Seed Corn or Prime Water	21
3. Too Poor to Tell	27
4. A Terrific Howl	35
5. Skimming Government Cream	40
6. Bound for the Promised Land	47
7. How the Cow Ate the Cabbage	55
8. Life Without Santa Claus	61
9. An Opportunity Within Reach	69
10. Across a Frontier	77
11. Principles of Cooperation	83
12. Big River Blues	90

13. Between Two Factions	96
14. A Dog Named Jake	102
15. Good Days and Bad Days	111
16. A Very Sad Occasion	117
Epilogue: A Poor Man's Best Proposition	123
Colony Credo	126
Appendices	
Appendix A: Glossary of New Deal Agencies and Programs	*128*
Appendix B: Questionnaire for Prospective Rural Colonists	*130*
Appendix C: Caseworker Analysis of a Prospective Colonist	*132*
Appendix D: Applicant Interview	*135*
Appendix E: Colony Officials and Roster of Colonists by County of Origin as of May 1, 1936	*143*
Appendix F: Homestead Sales Contracts	*149*
Appendix G: Colonists Who Moved Away	*157*
Appendix H: Memories of a Lifetime Participants Cited	*159*
Selected Bibliography	163
Index	175

Notes and Acknowledgments

The genesis of this edition is a slender volume that I published during 2015 in conjunction with the opening of Historic Dyess Colony: Johnny Cash Boyhood Home heritage site. Since that time many former colonists and their descendants have donated numerous official documents, letters, and photographs to the Dyess Colony archives. Additionally, more than 50 oral histories have been collected as part of an ongoing Memories of a Lifetime project. This book draws on these new resources to present a more expansive telling of the Dyess story.

Unfortunately for the author and reader, a colony history requires references to what seems like a bewildering array of government agencies. But it cannot be otherwise and be complete. During the Great Depression President Franklin D. Roosevelt's administration often launched new departments and programs. Several had major impacts, direct and indirect, on Dyess Colony. Acronyms are used in this text after first references to these organizations to save space, and an appendix provides a list of them for readers who become hopelessly confused.

When writing about Dyess there is an ongoing temptation to turn the story into another account of the life of J. R. (Johnny) Cash, the colony's most famous resident. However, several biographies of the international music star already exist and are noted in the bibliography. So references herein to Cash family members are related to events at the colony. Given that several are mentioned more than once, I have chosen to use their first names after my initial introduction in order to clarify identities.

Notes and Acknowledgments

Though the Dyess experiment has been called many things, including an example of socialism sweeping the land, here it is described as an agricultural resettlement colony, or colony for short. This seems to be the least controversial definition since within its boundaries at one time or another there existed cooperatives, sole proprietorships, joint ventures, and corporations.

The term "tenant" is used frequently for persons who rented land to make a crop. Tenants usually fell into four categories. Cash rent tenants paid a fixed sum of cash for the right to farm property, usually at the beginning of the year or end of the year. Crop share tenants paid the landlord a percentage of the crop as rent, such as 25 percent of cotton harvested. A hybrid crop share/cash lease possessed features of both forms. In these three rent arrangements a tenant usually provided all of the equipment and inputs necessary to make a crop, and farmers operated relatively independent of landowners. Sharecroppers, on the other hand, possessed few or none of the assets necessary to produce a crop, including cash, credit, equipment, or land. In this category landlords provided all or most of the necessary inputs except labor. For such extensive participation they charged a higher rent, often 50 percent of the crops.

Several frequently used words in this book come from a rural southern lexicon that may be unfamiliar to general readers. Tenants often are said to be renting from "planters," a term carried over from the South's plantation economy. Used here the word means landowners. Another term found in the text is "furnish." It is credit provided by suppliers to tenants for purchase of farm inputs and personal items. The accounts usually had to be paid when crops were sold in the fall. The word "gumbo" refers to a black soil that can stick like tar when wet and form chunks sometimes as big and hard as brickbats when dry. "Doodlum" means a form of currency typically issued to tenants in booklets for purchases in a plantation store. It could be a form of furnish. "Shotgun houses" tended to be the common abode of tenant farmers and laborers. These were long, narrow structures with front and back doors aligned so that if you shot through the front door the bullet would pass through the house and out the back door without hitting anything.

Those persons involved with Dyess sometimes referred to the "center" or "community center." In some cases they appear to mean a community building. In other uses they seem to be referring to a cluster of structures at the center of the colony that housed a commissary, café, shops, and administrative offices. The author takes such references to mean buildings at the colony's geographic center.

The stretch of land popularly known as the Mississippi Delta extends from Illinois to the Gulf of Mexico, but my focus is a part of that region called the Arkansas Delta. This strip of flat, fertile ground west of the Mississippi River begins in the state's northeast corner and ends in the southeast corner. References throughout this book to the Delta refer to the Arkansas portion.

An examination of materials related to Dyess reveals discrepancies as to the acreage purchased to establish the project and average price per acre. Official documents indicate that the colony consisted of about 16,000 acres acquired by the government at an average price of $2.50 per acre. The ultimate cost to colonists, however, ran considerably higher after factoring in development expenses such as clearing land, building houses and outbuildings, establishing community services, and installing roads, drainage ditches, bridges, and utilities.

The resettlement colony in Mississippi County first was named Colonization Project No. 1 but became known as Dyess Colony in honor of its founder, William R. Dyess. He died in an airplane crash on January 14, 1936. Though Dyess remains an Arkansas community my focus is from its origins in the early 1930s to its status in 1945. To some extent this is an arbitrary timeline, but in many ways Dyess no longer qualified as a resettlement colony by the end of World War II. It had become a small farm town and beyond the scope of this book.

Citations include footnotes at the bottom of each page. When referenced in illustrations, the source is noted at the end of each caption. Many oral histories included came from Memories of a

Lifetime, a research program at Arkansas State University to collect Dyess stories. Interviewers have included Mike Bowman, Ed Salo, Ben Manatt, and students in the Bentonville High School EAST (Education Accelerated by Service and Technology) Initiative. These are cited as "Memories of a Lifetime" with names of persons interviewed and date of interview. Newspaper citations are given at point of use by date of edition and name of the newspaper so they are not included in footnotes or the selected bibliography. Appendices use documents to outline stages in the resettlement process.

Several knowledgeable persons were kind enough to read the initial manuscript and provide sage advice: Michael B. Dougan, Distinguished Professor of History Emeritus at Arkansas State University; Everett Henson, a Dyess alumnus and keeper of the flame, and Joanne Cash Yates, Johnny's sister, born and raised at Dyess. My wife, Dr. Ruth Hawkins, led restoration of the Cash home and development of a colony museum as an Arkansas State University heritage site. As is true with all my books, her assistance and encouragement proved incalculable.

Dyess officials recruited out-of-work farm families from all 75 counties in Arkansas. They relocated to Mississippi County in the far northeastern corner of the state for a new start at what became Dyess Colony.

Prologue

In March 1935 a government truck carried Ray Cash's family from Cleveland County, Arkansas, to their new home in Dyess, an agricultural resettlement community located in Northeast Arkansas and financed with federal dollars. The community existed for relocation of struggling tenant farmers and laborers from throughout the state to a place where they could acquire a home and farmland at reasonable prices with favorable loan terms. Johnny Cash remembered his mother's reaction as the truck carried them farther and farther down muddy roads. "Sometimes Mama would cry and sometimes she'd sing, and sometimes it was hard to tell which was which."[1] The Cash children tried to sleep in the bed of this bouncing truck, but a tarpaulin used for cover did not protect them from frigid temperatures and freezing rain. What the Cash family discovered after their grueling journey, and what followed for them and families who joined them, is the story told here.

---

[1] Michael Streissguth, *Johnny Cash. The Biography* (Cambridge, Mass: Da Capo Press, 2006), 11.

Wilderness near Dyess, typical of colony land before clearing. Photo by Curtis Duncan

1
A STRONG BROWN GOD'S CREATION

To fully appreciate how Dyess came to be, one must first understand a place called the Delta, and this requires a trip back in time. Approximately 10,000 years ago the Mississippi River began to carry its present volume of water south to the Gulf of Mexico. Along the way it deposited rich alluvial soil that became a flood plain called the Mississippi Delta. It begins in Illinois and stretches to the Gulf of Mexico. During these thousands of years this mighty river, which poet T. S. Eliot called a "strong brown god,"[2] did violence to land along its banks despite efforts of men to harness the river for their use. The Mississippi devoured nearby earth and entire towns, sweeping them away and constantly creating new paths southward. Its water left many Delta acres an almost impassable wilderness, and for thousands of years the region remained that way. Contour of the land and its thick and jumbled ground cover resulted to some extent from major earthquakes centered near Northeast Arkansas that occurred in 1811 and 1812. They left ground covered with water, fallen trees, and crevasses. During his post-quake trip through the region naturalist Thomas Nuttall verified this foreboding landscape. He described a Delta "mostly low and clothed with enswamped forests or dense thickets of shrubs or canebrakes, monotonous and dreary, and uninhabited due to annual inundations."[3]

---

[2] T. S. Eliot, "Dry Salvages," *Four Quartets* (New York: Harcourt Brace, 1943), 21.
[3] Jeannette Graustein, *Thomas Nuttall Naturalist: Explorations in America 1808-1841* (Cambridge, Mass: Harvard University Press, 1967), 137-138.

When settlers arrived they gradually pushed aside Native Americans by any means necessary and began clearing this land. Many small farmers lived on Crowley's Ridge, high ground that runs from Southeast Missouri down to Helena, Arkansas. They typically engaged in mixed farming and orchards. During the first half of the 19th Century men from states such as Kentucky and Tennessee brought enslaved persons to convert the Delta's vast wetlands into arable acres. Beneath a scalding sun, often in fetid water up to their hips, and surrounded by swarming mosquitoes, enslaved laborers gradually drained and cleared sections of this land. They cut down trees, saved good wood for construction, and piled the remainder with brush for burning. After fires consumed the vestiges of swamp, new ground stood ready for cultivation. Though plows soon turned over Delta earth in much of the cleared ground, most acreage that would become Dyess remained wilderness until the 1930s.

Large sections of this region grew cotton, a crop that brought much money and misery to the South. The plant is a shrub named gossypium. Usually standing no higher than a typical man's shoulders, it became the Delta's primary crop and a profitable export. Growing and harvesting cotton in the 1800s and early 1900s depended on hand labor. After emerging from the ground, rows of the plants had to be blocked — a process of leaving hardy plants about one foot apart — and chopped to remove grass and weeds. When the plants became stalks, blooms appeared and later turned into hard green bolls about the size of golf balls. Bolls opened in the fall revealing fluffy cotton to be picked from the boll by people crawling down rows and dragging sacks several feet long. Hand labor continued into the first half of the 20th Century until mechanical pickers dominated the task. Cotton prospers in this region because of rich soil and an ideal growing season, usually lasting from mid-March to the end of October. With a mean temperature averaging about 60 degrees and ample rainfall, the Delta usually yields bountiful harvests.

Land sales agents promised wonderful harvests to potential buyers during the 1830s, and immigration increased when Arkansas joined the United States in July 1836. Many new arrivals depended on enslaved labor to make crops, and that peculiar institution helped set

Arkansas and other southern states on a course leading to a ruinous Civil War. At the end of that conflict, with much of the Delta in ruins, loss of enslaved labor sent southern planters on a search for vulnerable workers. They targeted freedmen and their families. Though now free, after the war many black workers labored under restrictions that previously governed slaves. Delta planters regained control of a black workforce after federal authorities grew weary of protecting freedmen. Many patterns between landlords and farm workers established during this period would be common throughout the South when Dyess became a reality.

Planters instituted onerous practices. A typical arrangement required hired help to begin work at daylight, take less than an hour for lunch, and work until dark. This was referred to by some weary workers as "from can to can't" (from when you can see in the morning until you can't see at night). Hungry men with their families in tow often faced a take-it-or-leave-it proposition, and they had to take it. To avoid paying cash wages, landlords promoted the concept of sharecropping, and this arrangement became widespread throughout the region. A February 2, 1867, *Little Rock Daily Conservative* editorialized that Arkansas agricultural leaders "conceded that the system of planting on shares is the most successful, as well as the most advantageous to both planter and laborer." It certainly proved advantageous for most planters, but rarely for laborers. Many landlords rigged results of tenant shares by charging exorbitant prices for crop inputs advanced to farmers and cheating them out of crop sale proceeds. The prevailing plantation system developed ways to keep a source of cheap, politically powerless workers.

This treatment brought bitter reflections from Arkansan Henry Blake, a freed slave and tenant farmer, during a 1930s interview. "After freedom, we worked on shares a while. Then we rented. When we worked on shares, we couldn't make nothing—just overalls and something to eat. Half went to the other man, and you would destroy your half if you weren't careful. A man that didn't know how to count would always lose. He might lose anyhow. They didn't give no itemized statement. No, you just had to take their word. They never give you no details. They just say you owe so much." Blake explained that during those Delta days and many that followed, tenants could not stop the cheating. "Brother, I'm telling you the truth about this. It's been that way for a long time. You had to take the white man's word on notes and everything. Anything you wanted, you could git if you were a good hand. You could git anything you wanted as long as you worked. But you better not leave him. You better not try to leave and git caught."4 This form of peonage eventually would be used to dominate poor whites as well.

Dreadful labor conditions existed throughout Mississippi County, Arkansas, which would become the home of Dyess Colony. Formation of the county came about in November 1833 after the territorial legislature combined land from three other counties. An 1840 census indicated that Mississippi County's population consisted of about 900 whites and 510 slaves. Author S. E. Simonson describes the county's terrain as "one alluvial plane [sic] without a hill or rock in the county. The rich alluvial soil is hundreds of feet in depth."5 Unfortunately, in the early 1800s several feet of water covered much of this rich soil. Large sections of the county remained a swampy wilderness until the early 1900s when several prominent planters and businessmen helped institute a drainage program. Simonson claims that turning swamps into arable ground helped increase the county's agricultural production by several hundred percent. It also increased values of real estate owned by planters. Despite such improvements, author Nan Elizabeth Woodruff points out that "the Arkansas Delta retained a frontier quality well into World War I. Sections of the region had larger percentages of white sharecroppers and tenants than did those on the Mississippi side."6 This point is validated by a 1910 census indicating that blacks made up only 4.2 percent of the county's population.

---

4 Henry Blake, Little Rock, Arkansas. Interviewed by Samuel S. Taylor, "Born in Slavery: Slave Narratives from the Federal Writers Project, 1936-1938." Arkansas Narratives, 2:1, Manuscript Division, Library of Congress.

5 S. E. Simonson, "Origin of Drainage Projects in Mississippi County," *Arkansas Historical Quarterly*, 5:3 (Autumn 1946), 264-265.

6 Nan Elizabeth Woodruff, *American Congo. The African American Freedom Struggle in the Delta* (Cambridge, Mass: Harvard University Press, 2003), 30-31.

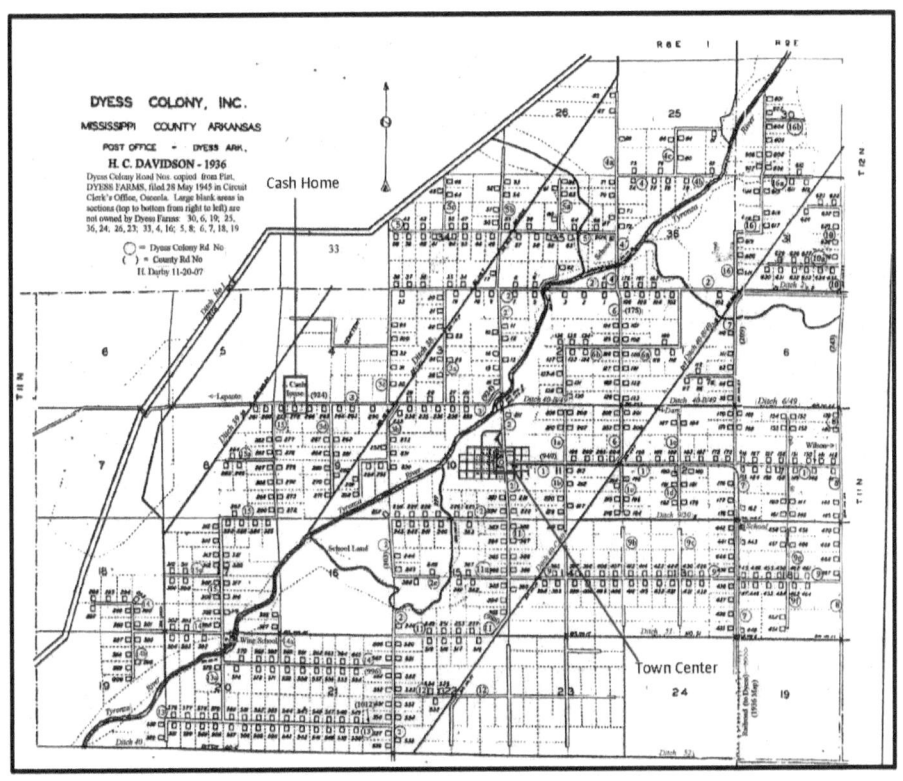

Some 16,000 acres of forested, swampy land was converted into what became Dyess Colony after draining swamps, building roads and bridges, and staking out and developing farmsteads (primarily 20 acres each.) National Archives. Records of the Work Projects Administration

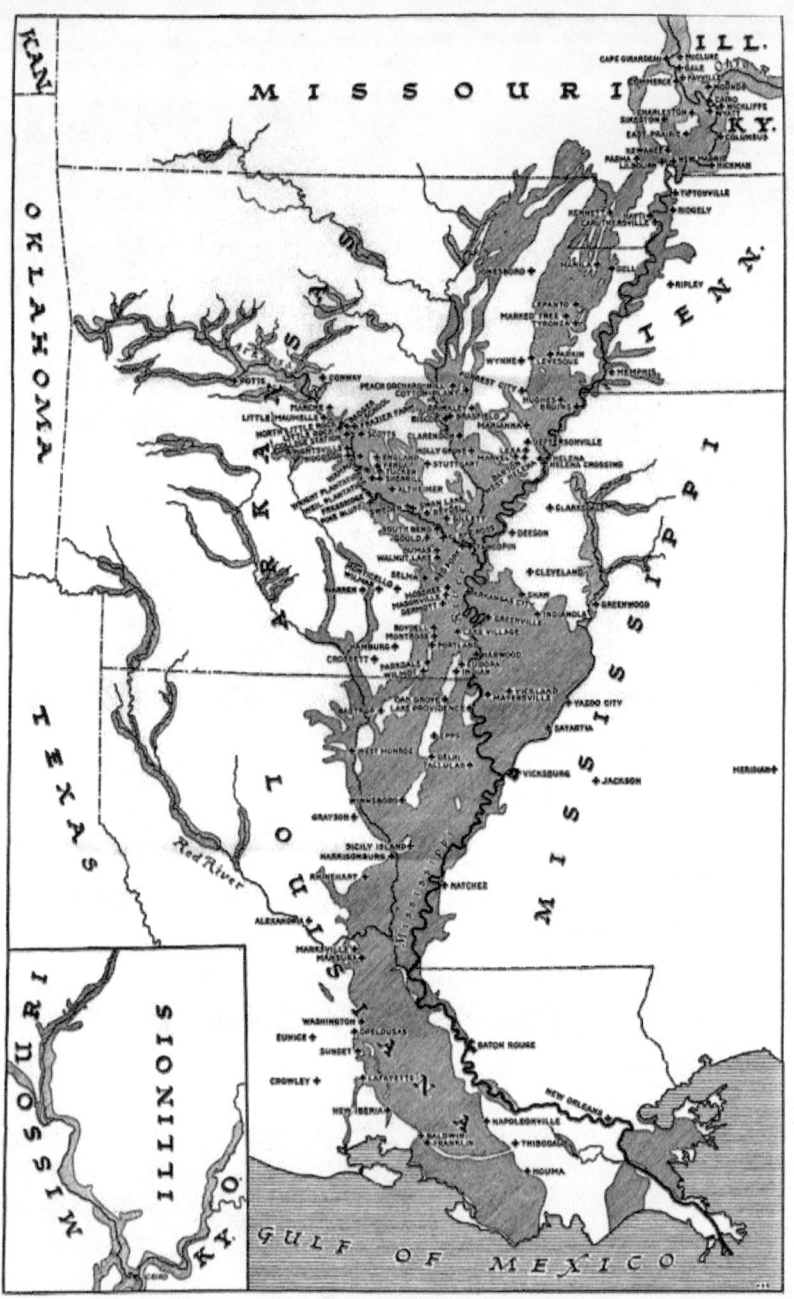
Dark areas show the 1927 Mississippi River flood reaching all the way to the western borders of Arkansas, thanks to tributary overflows. American Red Cross

2
NO SEED CORN
OR PRIME WATER

Though American farmers learn early in life never to eat their seed corn or drink the pump prime water, many did both to sustain themselves during the 1920s. Hard times for sharecroppers began in 1919, not 1929. President Herbert Hoover, whose name forever would be associated with a failure of leadership during the early 1930s, pointed to events well before the Depression decade as reasons for the financial collapse. He writes in his memoirs that economic "disturbances have many roots in the dislocations from the World War."[7] Nor was he alone in this belief. Author Sidney Baldwin agrees. "For agriculture as a whole, the Great Depression began not on the fateful day in October 1929, but in 1920, when farm commodity prices suddenly collapsed and the war-time boom dissolved."[8]

That war began when the guns of August roared in Europe during 1914. It devastated confidence in American commodity markets, and that fall major cotton exchanges failed to open in New Orleans, Memphis, and Galveston. "People are losing their heads," one observer told the U. S. Senate. The price of cotton went down about $10 a bale, and cotton exchanges closed for three months. With no reliable barometer of prices, farmers were exploited by spot buyers offering eight cents or less. "Some found their crop unmarketable at any price,

---

[7] Herbert Hoover, *Memoirs of Herbert Hoover: The Great Depression, 1929-1941* (New York: Macmillan, 1952), 105.
[8] Sidney Baldwin, *Poverty and Politics. The Rise and Decline of the Farm Security Administration* (Chapel Hill: University of North Carolina Press, 1968), 32.

hauled it to town, and then back home again."⁹ Many state governments considered actions to ease pressure on producers. One approach proposed a limitation on cotton acres planted to reduce supply of the fiber, but President Woodrow Wilson rejected the plan. A federal policy that required plowing up cotton and reducing future acreage planted would resurface during the New Deal and be adopted nationally. It helped give rise to Dyess and other programs to assist displaced tenants.

Fortunately for American farmers, financial opportunities improved when the United States entered the war in April 1917. Industry expansion to meet demand included southern textile mills, and some enterprises reported growth in excess of 100 percent. According to author George Brown Tindall, "The taste of prosperity was sweetest of all to the subjects of King Cotton."¹⁰ The price jumped to 35 cents per pound. In addition to use of cotton in making many thousands of uniforms, manufacturers bought the fiber to help produce explosives. Generally favorable prices continued throughout the war, and in 1919 many tenants used this boom to embark on a buying spree. Some southern farmers bought $45 suits (about $1,000 in 2020) and had the audacity to purchase an automobile. "They are like little children to whom some good fairy has paid a visit," wrote the *Savannah Press*, ignoring the fact that for many poor farmers it was the first suit they had ever owned. "Our people can't stand prosperity," an Arkansan asserted.¹¹ Prosperity soon ended with severe price declines after a 1918 armistice. A huge 1919 crop harvested at the end of that year led to plunging prices in 1920. A report prepared for the federal agriculture secretary in 1921 pointed to the second largest harvest in American history. This report indicated that the country's ten primary crops had combined yields 13 percent above averages for five years preceding outbreak of the war.¹²

---

[9] George Brown Tindall, *The Emergence of the New South 1913-1945*. (Baton Rouge: Louisiana State University Press, 1967), 33.
[10] Ibid., 60.
[11] Ibid., 61.
[12] Wayne D. Rasmussen, *Agriculture in the United States. A Documentary History* (New York: Random House, 1975), 2645.

Unfortunately for most farmers the cost of crop inputs remained inflated as crop prices deteriorated. Cyclical cotton prices slammed the market sharply up and down throughout the early 1920s. During those unstable years good times were not good enough to offset bad times, which led to foreclosures and bankruptcies throughout rural regions. Perennial problems with agriculture produced many experimental ideas, some of which reappeared at Dyess. By 1926 several states had created land settlement boards that purchased land, erected farm buildings, put in irrigation systems, and made homesteads available at reasonable prices and terms. State leaders hoped to lure families back to the farms. At that time one producer could handle only about 34 acres. Those with machinery might increase the operation to 100 acres. So vast expanses of arable land lay idle throughout the country while a hungry population increased. One idea for price supports came out of a Memphis, Tennessee, cotton convention during October 1926. This concept required bankers to refuse loans to farmers unless they cut their cotton acreage by 25 percent. A reduced cotton crop in exchange for government cash would become a centerpiece of President Roosevelt's Agricultural Adjustment Act (AAA) implemented in the 1930s.

Adding to regional woes, a record-setting 1927 flood followed heavy rains that started near the end of 1926 and continued into spring 1927. Water covered states with tributaries flowing into the Mississippi River, and it all surged southward. An April 15, 1927, *Memphis Commercial Appeal* issued an ominous warning. "The roaring Mississippi River is believed to be on its mightiest rampage. All along the Mississippi considerable fear is felt over the prospects for the greatest flood in history." It was indeed. According to author John M. Barry, "The river seemed to be the most powerful thing in the world. From the breadth of the continent down had come all the water that fell upon the earth and was not evaporated into the air or absorbed by the soil, down as if poured through a funnel, down into this immense writhing snake of a river, this Mississippi."[13] The raging river first overcame a regional levee in April, and during following weeks water

---

[13] John M. Barry. *Rising Tide. The Mississippi Flood of 1927 and How It Changed America* (New York: Simon and Schuster, 1997), 16.

rushed through other cracks in levees and flooded about 25,000 square miles of the Lower Mississippi Valley. An observer called it "a vast sheet of water as yellow as the China Sea . . . 1,050 miles long and in places over 50 miles in width."[14] The flood brought varying degrees of ruin to 200 counties in eight states. Only 127 Arkansans lost their lives, but approximately 40,000 had to abandon their homes. The flood covered about 13 percent of the state, and the total cost to Arkansas was estimated to be $15 million.

Cognizant of farm woes caused by man and nature, federal officials considered various relief measures, including one resembling what would occur at Dyess. It originated in the U. S. Department of the Interior. An official from that agency during December 1927 gathered delegates in Washington, D. C. from nine southern states. This group drafted a bill to establish and provide government support for rural communities in 12 states. The bill never became law, but it helped lead the way for New Deal rural communities established a few years later.[15] Many ideas advanced to help farmers in the 1920s failed for the most part. A South Carolina planter said that price declines contributed to "permanent scars. The South had lost its first opportunity since the Civil War to accumulate capital and break the chain" of farmers in hock to merchants and bankers.[16]

Clearly the "Roaring '20s" raced past many agricultural regions. Author David Kennedy agrees that "a severe economic crisis beset the farm belt almost as soon as the Great War concluded."[17] Though farm income and purchasing power declined in the decade, wages of industrial workers increased by almost 25 percent. Despite the nation's general prosperity tremors began to shake the confidence of some commercial enterprises. Real estate values softened in areas such as Florida, and business inventories increased. The stock market

---

[14] Bruce A. Lohof, editor, "Herbert Hoover's Mississippi Valley Land Reform Memorandum: A Document," *Arkansas Historical Quarterly*, 29: 4 (Winter 1970), 112.
[15] Tindall, 131.
[16] Ibid., 112.
[17] David M. Kennedy, *The American People in the Great Depression. Freedom From Fear. Part One* (New York: Oxford University Press, 1999), Preface, x.

had sustained a frenzied pace caused by what one might call irrational exuberance, but in fall 1929 the economic bubbles began to burst. On Wednesday, October 23, a flood of sell orders dumped six million shares of common stock and eliminated $4 billion of market value. Weeks of losses wiped out gamblers big and small. The tab for October-November reversals came to about $26 billion, which equaled one-third of the market's total value before the collapse began.

Investment losses wiped out capital necessary to carry on commerce, so businesses began to fail. This put both owners and employees out of work and removed a market for other goods and services. On and on it went, each collapsing domino overturning the next in line. Downward momentum continued as the new decade began, and bank failures increased. Many undercapitalized rural banks went under first, and their customers sank with them. Most farmers made their crops with borrowed money, and when there was no money to borrow there was no crop to make. Some growers who managed to put in a crop often found no market for it and abandoned it in the field. Food crops rotted while hungry people in cities scavenged for meals, a perplexing irony. Unfortunately, Hoover lacked the philosophy, programs, and political skill to cope with it.

Failure punished almost every living creature, one way or an other. The unemployment rate hit 17 percent, but even those employed often lived in appalling conditions. Many agricultural laborers survived on rations no better than those fed to farm animals. Horror stories abounded. An observer saw a crowd of about 50 men fighting over a barrel of garbage that had been set outside the back of a restaurant. In one rural school the teacher told a child who looked sick to go home and get something to eat. "'I can't,' the girl replied. 'It's my sister's turn to eat.'"[18] During the 1920s some state legislators called Arkansas the "Wonder State." After that decade ended they may have wondered what happened to their state. Most rural Arkansans, already broke, had no nest egg to draw upon, and the chickens were coming home to roost.

---

[18] T. H. Watkins, *The Great Depression* (New York: Backside Inc., 1993), 57.

The 1930-31 drought led to much suffering and malnutrition, including this child with rickets in Mississippi County, Arkansas. Library of Congress, Photo by Arthur Rothstein

3
Too Poor
to Tell

In his "Song of the South," Woody Guthrie wrote this about the Depression: "Well, somebody told us Wall Street fell, but we were so poor that we couldn't tell." His claim applied to many Arkansans. Author William D. Downs Jr. points out that in 1929 the state ranked 46 in per capita income and was broke.[19] So were its citizens. The situation became particularly acute in rural areas. A 1934 study of 646 cotton plantations pegged an average net income at only $312 a year per family and $71 a year per person for sharecroppers.[20] The 1930s began with a record-shattering drought in several southern states, including Arkansas, that added to difficulties. According to an August 12, 1930, *Arkansas Gazette*, the National Weather Service declared conditions in the region "the most severe drought in climatological history." Summer temperatures hovered around 110. Fish poached in ponds, and garden greens wilted. As June and July passed into August, Little Rock recorded its 71st day without precipitation. Arkansas congressmen sought help from Hoover, but the President had a narrow view of federal intervention. "It is not the function of the government to relieve individuals of their responsibilities to their neighbors, or to relieve private institutions of their responsibilities to the public, or of local government to the states, or of state government to the federal government."[21]

---

[19] William D. Downs Jr., *Stories of Survival. Arkansas Farmers During the Great Depression* (Fayetteville, Ark: Phoenix International, Inc., 2011), 19.

[20] Anna Rochester, *Why Farmers Are Poor* (New York: International Publishers, 1940), 60.

[21] Broadus Mitchell, *The Depression Decade* (Armonk, N.Y.: E. Sharpe, 1975), 87.

A NEW DEAL IN DYESS

The President insisted that help for those suffering from the drought should come primarily from charitable organizations such as the Red Cross, and Arkansas became the first major recipient of Red Cross support.[22] A Mississippi County aid worker received 1,200 requests for help in just two days. Another agent explained the insidious nature of this crisis in a January 9, 1931, *Arkansas Gazette*. "Unlike a spectacular flood or cyclone, starvation has crept up so slowly that people are unaware of the dire need and destitution. People have no food and no clothing." Red Cross officials aimed to provide needy families a $2.00 per week food allowance. But the total for Arkansas families came closer to $1.20 per month, which kept them one short step away from starvation. Public dissatisfaction with Washington's response finally prodded the U. S. Congress into action. A Federal Drought Relief Act of 1930 enabled the secretary of agriculture to make loans to farmers victimized by bad weather, but they came with dreadful restrictions. Proceeds could not be spent to purchase food and clothing for farm families. The money had to be used exclusively for seed and other crop inputs needed in the spring. So some drought-stricken farmers faced a winter with credit available, but useless in helping them feed their families and heat their homes.

As failures accumulated, editorial writers and Democratic Party operatives anchored the disaster to Hoover, and it sank his political career. His Presidency lasted only one term. In May 1933 after Roosevelt's inauguration that March he gave a speech at Oglethorpe University that signaled steps he would take to prop up a collapsing economy. The new President made this commitment: "The country needs and, unless I mistake its temper, the country demands bold, persistent experimentation. It is common sense to take a method and try it: if it fails, admit it frankly and try another. But above all, try something."[23]

As it turned out, Roosevelt tried almost everything in an effort to find something that would work. At the time of his election triumph

---

[22] Roger Lambert, "Hoover and the Red Cross in the Arkansas Drought of 1930," *Arkansas Historical Quarterly*, 29:1 (Spring 1970), 3.
[23] Kennedy, 104.

about 13 million workers lacked jobs, and every fourth household in America had no breadwinner. Many Americans feared that they were witnessing not just a massive market downturn, "but the collapse of an historic economic, political, and social order, of the entire American way of life."[24] The new President began to attack this lack of confidence during the first 100 days of his administration. He created new departments to aid the needy, including some to assist farmers. The administration established four agencies that commenced working with poor people: a Division of Subsistence Homesteads in the U. S. Department of the Interior; a Division of Rural Rehabilitation (RR) in the Federal Emergency Relief Administration (FERA); a Resettlement Administration (RA), and a successor agency, the Farm Security Administration (FSA).

FERA would become instrumental in creation and support of Dyess Colony. A field service supervisor explained that agency's philosophy. "Rural rehabilitation is not a program of public charity. Beneficiaries are expected to give their notes for repayment in full for both subsistence and capital goods advances. There are no reservations regarding the matter of repayment. Every promise to pay is assumed to be made in good faith. Indications are that rehabilitation clients themselves favor this policy."[25] Through FERA's efforts destitute families on relief could purchase farmland with a 30-year loan amortization. Works Progress Administration (WPA), created to put the unemployed to work on public service projects, paid for houses, roads, and ditches for communities. Headed up by Rexford Tugwell, the RA coordinated assistance to impoverished country folks.

Despite obvious desperation in the Delta, government assistance galled some observers. Federal programs drew fire from a May 25, 1933, *Memphis Commercial Appeal*. It claimed that federal agencies had "fallen into the hands and under the blight of social gainers, do-gooders, bleeding-hearts, and long-hairs who make a

---

[24] Ibid., xi.
[25] "Rehabilitation Policies," Summarized by Paul V. Maris, Supervisor of Field Services, Arkansas Federal Emergency Relief Administration. Everett Henson Collection, Dyess Colony Archives #2016-33-010, Arkansas State University.

career of helping others for a price and according to their own peculiar, screwball ideas." FERA's public assistance efforts also developed critics on the other end of the political spectrum. Harry Hopkins, Roosevelt's close aide, complained that disrepute attached to those seeking public assistance came from a belief that an applicant must "in some way be morally deficient. He must be made to feel his pauperism. Every help which was given him was to be given in a way to intensify his sense of shame."[26] Sometimes derision came from those charged with providing FERA assistance, and it could be glaringly racist. A southern relief director said, "Any nigger who gets over $8 a week is a spoiled nigger."[27]

Of all New Deal programs meant to help farmers, the Agricultural Adjustment Act (AAA) probably did the most to impoverish southern tenants. When Roosevelt appointed Henry A. Wallace secretary of agriculture, the price of cotton languished at six cents per pound. Almost 13 million unwanted bales filled warehouses. The President's advisors conceived a program to increase commodity prices, and it did somewhat. It also validated the law of unintended consequences after Roosevelt signed the AAA in May 1933. The theory behind this law appeared to be unassailable. A reduction of acres planted to a crop should reduce inventory and thus increase prices. The federal government would give farmers cash payments to compensate them for lost income. The program called for a 25 percent reduction of Arkansas cotton acres, the bulk of it in the Delta. Tenants supposedly would share in the payments pro rata. Unfortunately for them, AAA checks came to landowners, and many kept all the money for themselves. Since less acres required fewer tenants, often planters threw them out of their homes, off the land, and made day laborers out of a few who remained. Direct governance of the program fell upon county extension service agents. They selected members to serve on production committees with control over planted acres and

---

[26] Harry Hopkins, *Spending to Save: The Complete Story of Relief* (New York: Norton, 1936), 100.
[27] Kennedy, 173.

government money. Often these committees consisted of planters and business associates.

When protesting landlord abuses, tenants often found that they had in actuality no recourse for ill treatment. Section 7 of AAA regulations stated: "The producer will endeavor . . . to bring about reduction as to cause the least possible . . . social disturbance and . . . insofar as possible he shall effect the acreage reduction as nearly ratable as practicable among tenants . . . [and] shall insofar as possible maintain . . . the normal number of tenants." Many planters used this porous language to cheat desperate people out of their share of government money and turn them into day laborers or vagabonds.[28] Displaced tenants who complained about program abuses received little sympathy or assistance. Politics caused the problem, and it went all the way to the top. Though Roosevelt claimed to be offering a new deal to poor farmers his secretary of agriculture assured prominent planters that federal regulations would meet "the wishes of the farm leaders conference."[29] Simply put, the President needed votes of southern senators and representatives to pass his political agenda. They needed support and money from constituent landowners, and apparently few needed the support of tenant farmers. Social justice activist Howard Kester looked into abuses and reached an alarming conclusion. "AAA policies of the federal government intensified the already deepening misery of the southern sharecropper."[30]

During the early 1930s Arkansas Congressman Brooks Hays undertook a fact-finding mission for FERA that took him down many dirt roads in the Delta. His shocking report painted a dismal picture of life there. "I had seen many dilapidated houses, but I could hardly believe that people lived in the one that I found on this country road. This home was about ten feet by twenty, and was made of corrugated tin and scraps of lumber. It was flat upon the ground, and had only one

---

[28] Howard Kester, *Revolt Among the Sharecroppers* (New York: Arno Press, 1969), 30.

[29] Henry A. Wallace, *New Frontiers* (New York: Reynal & Hitchcock, 1934), 164-165.

[30] Kester, 54.

or two small openings."[31] For many years Hays excoriated those who stripped the poor of their earnings and dignity, and he blamed deplorable conditions among Arkansas tenant farmers on absentee land ownership. His solutions included turning enterprising tenants into independent farmers. Dr. Carl Taylor, director of RA's rural resettlement division, shared this view of tenancy. Taylor believed that adequate rural life could not be achieved with farm tenancies. Most, if not all, who became Dyess farmers agreed with him.

---

[31] Donald Grubbs, *Cry From the Cotton: The Southern Tenant Farmers Union and the New Deal* (Chapel Hill: University of North Carolina Press, 1971), 5.

Typical houses provided for sharecroppers and tenant farmers were makeshift buildings, pieced together from scrap lumber with no plumbing, electricity or window screens.
Photo by Curtis Duncan

Above, a crowd of angry depositors waits in the rain to withdraw money from the Bank of the United States after its failure in 1931. Library of Congress. *Below, hard times in the Arkansas Delta led to creation of the racially integrated Southern Tenant Farmers Union in 1934.* Photo by Louise Boyle, STFU Records, Southern Historical Collection, Wilson Library, University of North Carolina, Chapel Hill

4
A TERRIFIC HOWL

When the 1930s began, Arkansas' unemployment rate reached 39 percent. Several events made difficulties in the state especially severe: the Wall Street crash and its effects on the national economy; natural disasters such as a massive flood and severe drought; depressed farm commodity prices, and failing banks causing a lack of access to credit. Eventually more than 100 Arkansas banks went out of business. Some of the state's destitute reportedly lived in caves, and at least one didn't live at all. A January 1, 1931, Pocahontas, Arkansas, *Star Herald* reported that a despondent Nimmons, Arkansas, farmer who had five children killed himself with a pistol. Though also despondent, more than a few Delta tenant farmers in shotgun houses felt lucky to have what little they possessed. Their work days began at dawn after families shared a breakfast often of biscuits and redeye gravy. On good days they had molasses and maybe fatback. Every member of the family went to the field, children included, and at sundown trudged back to their shacks. Various people who studied living conditions of farm labor reached similar conclusions. A federal inspector described the tenant farmer's life as a "picture of squalor, filth, and poverty.[32] Another government analyst characterized homes as "dilapidated, damp, or overcrowded, with lack of sunlight, air, piped-in water, or

---

[32] Louis Cantor, *Prologue to the Protest Movement: The Missouri Sharecropper Roadside Demonstration of 1939* (Durham, N. C.: Duke University Press, 1969), 14.

proper sanitation."³³ They were hot during summer and almost impossible to heat in winter.

Many poor farmers who hoped for acceptance at Dyess were locked in a state of servitude and sometimes starvation by economic conditions and a system devised and maintained by landowners and their cronies. Like freedmen after the Civil War, they survived on credit from the beginning of each year until settling up time when they hauled their harvested crops to a gin often owned by their landlord or his business associates. Planters usually required tenants to buy farm supplies from their plantation store, often at exorbitant prices with high interest rates to carry accounts. As a result of such abuses some despairing farm workers and tenants formed one of the most effective and lasting rebellions against planter tyranny. On a hot summer day in 1934, 18 men met in a small building on Fairview Plantation south of Tyronza in Northeast Arkansas and formed the Southern Tenant Farmers Union (STFU). This was not the first organization of its kind, nor would it be the last. The Grange, Agricultural Wheel, and Farmers Union among others sought improvements as well.

The STFU grew out of earlier informal gatherings of poor farmers to share their financial desperation. This racially integrated union, with some women and blacks in leadership positions, gathered hundreds of members in several states and existed for more than two decades. It called for strikes to increase farm wages and crop shares received by tenants. Membership carried great risks. H. L. Mitchell, co-founder of the union, relates a story told him by an observer visiting Arkansas. "Near Marked Tree, about four miles out, we came across a family, a mother with five children, a couple of young hogs, six chickens, two puppies in the children's arms, and a few scraps of furniture. They had just been dumped on the side of the road by a deputy sheriff. The father came walking down the road a little later. He

---

³³ E. L. Kirkpatrick, "Housing Aspects of Resettlement," *Annals of the American Academy of Political and Social Science,* 190, Current Developments in Housing (March 1937), 94.

had been beaten up a few days before, arrested and jailed on the nominally untrue charge of stealing a couple of eggs."[34]

Perhaps due in part to negative publicity generated by such mistreatment of poor farmers, by 1934 several New Deal programs began to gain traction in the region. Farm Credit Administration (FCA) refinanced mortgages and saved some families from foreclosures. Commodity Credit Corporation loans helped protect growers from ruinous cyclical price downturns. While such resources sought to diminish rural poverty, other federal agencies established for the nation's general wellbeing also made improvements in Arkansas. They included WPA's construction of 16 hospitals, 297 schools, almost 1,000 buildings, and thousands of miles of roads. Civilian Conservation Corps (CCC) completed other important projects. Its members lived in 106 camps throughout the state. Workers received $30 per month to help with mostly outdoor tasks such as parks and bridges.

Lorena Hickok became a keen observer of these and other government programs. Hopkins hired her to travel throughout the country compiling data about the effects that FERA and other federal agencies were having on the nation's massive economic problems. Hickok's stories sent to Hopkins in the form of letters sometimes revealed an astonishing level of insensitivity among the wealthy. In a letter to Hopkins during June 1934 she conveyed one Memphis cotton merchant's appalling attitude about poor southern farmers. "One wealthy cotton man and banker gives the impression that he thinks all tenants are lazy beggars and should be treated as serfs and would rather see the price of cotton stay down at 5 cents a pound forever than be boosted with government control."[35] In another letter to Hopkins she wrote: "The truth is that the rural South never has progressed beyond slave labor. When their slaves were taken away, they [white landowners] proceeded to establish a system of peonage that was as

---

[34] H. L. Mitchell, *Mean Things Happening in This Land: The Life and Times of H. L. Mitchell, Co-founder of the Southern Tenant Farmers Union* (Montclair, N. J.: Allanheld, Osmun & Co., 1979), 73.

[35] Richard Lowitt and Maurine Beasley, editors, *One Third of a Nation. Lorena Hickok Reports on the Great Depression* (Urbana: University of Illinois Press, 1981), 277.

close to slavery as it possibly could be and included whites as well as blacks." Planters at first supported federal government assistance, "but now, finding that CWA [Civil Works Administration] has taken up some of the labor surplus," they are panicky. Planters realize that they "may have to make better terms with tenants and pay day laborers more." They are "raising a terrific howl against CWA."[36] This mentality caused author H. C. Nixon to call the South an "economy of the Middle Ages without the cathedrals."[37]

---

[36] Ibid., 186-187.
[37] H. C. Nixon, *Forty Acres and Steel Mules* (Chapel Hill: University of North Carolina Press, 1938), 19.

Above, federal land surveyors begin staking out Dyess farmstead sites. Below, wilderness being cleared for housing at Dyess Colony. National Archives, Records of the Work Projects Administration

5
SKIMMING GOVERNMENT CREAM

Amid economic desolation and political turmoil a resettlement colony emerged in Northeast Arkansas that would provide a refuge for those fortunate enough to win acceptance there. Administrator of both FERA and WPA in Arkansas, William R. Dyess obtained federal funds to turn thousands of acres of Mississippi County wetlands into what he hoped would become a model agricultural community. Dyess first proposed the resettlement colony to Hopkins and his FERA assistant, Lawrence Westbrook. Hopkins considered helping rural residents to be a priority, noting that the "urgent necessity to care for the nation's rural and small town unemployed is shown by the fact that approximately 40 percent of the 5.1 million families on relief are to be found in the open country and towns under 5,000 in population."[38] Hopkins approved Dyess' plan and made funds available for purchase of land and development costs through the Arkansas branch of FERA.

Arkansas FERA established a separate corporation within its domain, Arkansas Rural Rehabilitation Corporation (ARRC). It became responsible for Colonization Project Number One, renamed Dyess Colony to honor its founder. Dyess bought approximately 16,000 acres of land in Township 11 and 12 North, Range 8 and 9 East, for an average cost of about $2.50 per acre. He purchased this Mississippi County swampland along the Tyronza River about 17 miles from the Mississippi River from three sellers: the plantation empire of Lee Wilson and Company, Creamery Package Company, and Drainage District Number Nine. FERA provided money to buy the land and

---

[38] Hopkins, Sept. 27, 1935, Hopkins press highlights, n.p., WPA Press Clippings, Box 5, Arkansas State Archives.

projected the colony's cost to be $1.5 million. Like many federal projects it would come in over budget—about one million dollars over by 1938. The deal drew criticism from the start. Author David Hayden asked an obvious question and provided three possible answers. Why did Dyess purchase unimproved swampland rather than ground already in production? (1) He had an interest in this area. He was from Luxora, a small town near the proposed colony site. (2) The land was cheap. (3) Arkansas Relief Administration could take men off relief rolls and put them to work improving this property.[39]

ARRC started with only three stockholders but added four more in 1934. After these additions, stockholders consisted of Dyess; Floyd Sharp, Dyess' assistant; Robert H. McNair Jr., chief of accounting at Arkansas FERA; Malcolm Miller, district field representative for Mississippi County FERA; Dan T. Gray, regional director of an AAA land policy section; T. Ray Reid, USDA assistant extension service director, and M. A. Wood, an ARRC member. These men possessed substantial political connections, and they would need them to fend off attacks by political enemies. Dyess' clout became formidable when in 1932 he supported J. M. Futrell for governor. Futrell won and advanced his ally with appointments to government relief programs that made available large sums of money to be spent in Arkansas for jobs and improvements. That money made Dyess a potent force. When Futrell ran for reelection Dyess used FERA resources to assist the campaign. In addition to political connections his charismatic personality proved to be a great asset. Gertrude S. Gates, a state relief program employee, admitted that Dyess "could charm the birds off the trees if he chose."[40]

Work at the colony formally began on May 22, 1934, when O. G. Norment led his first crew into that enormous wilderness. They made their way in by using a narrow trail. He later shared his first impression. "I shall never forget what a seemingly insurmountable task

---

[39] David Hayden, "A History of Dyess, Arkansas." Thesis Submitted in Partial Fulfillment of the Requirements for the Degree of Master of Arts. Department of History in the Graduate School, Southern Illinois University. August 1970, 14.

[40] Donald Holley, "Trouble in Paradise: Dyess Colony and Arkansas Politics," *Arkansas Historical Quarterly*, 32: 3 (Autumn 1973), 206.

we had before us. Practically the entire acreage consisted of cut-over hard wood timber land. There were a few cleared spots where the old logging camps had been established, most of these grown up, however, into a mass of bushes and small saplings." Norment and his men found some squatters in shacks on cleared land, but colony acreage for the most part had remained largely "uninhabited, inhospitable, and forbidding."[41] After completion of surveys, 115 logging mules and several hundred men went to work. Cone Murphy, executive administrator, led efforts to ready this land for habitation and cultivation. He hired skilled employees from Little Rock and trucked in common laborers from nearby towns. Many came from Mississippi County relief rolls. The total number of men employed to turn sections of the wilderness into a hospitable place would vary but peaked at about 1,500. This workforce became a significant source of employment for the county, with unskilled laborers earning $1.60 per day. Administrators and supervisors received monthly checks.

When lumber operations began, workers constructed six groundhog sawmills and one steam mill to convert fallen trees into lumber for use in project construction. Survey crews identified rights-of-way throughout the colony, and laborers eventually created 68 miles of gravel roads. Three draglines carved out ditches, and workers constructed 24 major bridges across streams. They deepened the Tyronza River channel and connected it to drainage ditches. A new five-mile rail spur between the colony and Hitt, Arkansas, allowed trains to bring in supplies. Sturdy, resilient men who accomplished all of this lived in military-like barracks with mess halls. They must have been a rowdy lot since police arrested an average of two each week for drunkenness. Consistent with the times, facilities segregated workers by race.

In June 1934, crews completed a temporary administration building to house staff members. Storage facilities and barns soon followed for use by construction teams. Workmen poured concrete for

---

[41] Mike Wade, "Founding of Dyess Colony," August 1979 manuscript chapter in possession of Jean Ann Cannon Jennings. Posted on Dyess Colony private website on June 21, 2005.

an administrative building and adjacent structures for commercial enterprises that would become the colony's community center. A Little Rock architect, Howard Eichenbaum, designed houses ranging from three to five rooms on farms with 20 to 40 acres. He used the South's famous shotgun house design to create some Dyess models but turned them sideways so that long sides became the front and back. Porches along the front and kitchen provided additional space, and each house size had seven or eight floor plans for variety. All contained electrical wiring and plumbing fixtures, though hookups were not available until nearly ten years after colonists first moved in.

Lack of electricity and plumbing led to improvisation. Ann Blue said her family would "heat water and put it in the bathtub. Of course when we were little kids the wash tub went in the living room by the fire and that's where we took our baths. I don't know when my parents took a bath. I wasn't around for that." She eventually had three brothers. "So the baths started out with my younger brother, and then I was second. I was second to the oldest, but I got the second bath because I was a girl. And you didn't get quite as dirty as those boys. And then it sort of went up the line."[42] Absence of electricity caused some troubles. Mary Lou Mauldin said that "Daddy had a radio and he was one of the few on our road that had one. But the battery was about as big as a car battery. And all the men around our neighborhood would come and gather around and hear Joe Lewis. And I remember the frequency was just on and off, on and off. They'd gather—put their ear right down to the radio. On one occasion the battery started smoking and Daddy knocked them all out of the way and threw the battery out in the front yard."[43] These homes may have been modest and lacked some conveniences, but when viewed in the context of

---

[42] Ann Roberts Blue, Oral History Interview, Memories of a Lifetime Project Team, October 20, 2017. Dyess Colony Archives #2018-12-010, Arkansas State University.

[43] Mary Lou Wilson Mauldin, Oral History Interview, Memories of a Lifetime Project Team, October 20, 2017. Dyess Colony Archives #2018-12-011, Arkansas State University.

tenant housing surrounding the colony they looked like "debutantes in the slums," according to newspaper reporter Jonathan Daniels.[44]

Though initial plans called for 600 houses to be constructed, revisions lowered that number to 500. While preoccupied with construction and other duties William Dyess became embroiled in accusations of crony capitalism. A July 6, 1934, *Osceola Times* asked several pointed questions. Why did the majority of relief labor come from Osceola? Why did Dyess locate the colony near his home? Why was much of the land purchased from a close friend's son, Lee Wilson Jr.? Did Dyess allow Wilson to sell worthless land to the government for a substantial profit? Why did Dyess purchase building materials and machinery for colony projects from his friend, Ben Butler? Why did government funds pay for gravel roads near Wilson land? After what appeared to be a relatively lenient investigation by FERA agents they found that Dyess committed an indiscretion by improving roads on land owned by Wilson, but no criminal acts occurred. As to all other matters, investigators found no unusual favoritism. Nevertheless, it appeared that Dyess' goal may have been less than altruistic and that he undertook the project "with his eye on skimming a ladle of government cream," as one author put it.[45] Regardless, remarkable achievements by colonists and positive reviews by most observers suggested that he achieved many worthy results.

Despite political sniping the project pressed ahead. Eichenbaum and David R. Williams, FERA's construction division chief, carved Dyess into 80-acre sections, which in turn were divided into four, 20-acre tracts. Their design took the shape of a wagon wheel with a community center as its hub. Roads had numbers, not names. The project eventually ended up with 61 three-room houses, 233 four-room models, and 206 five-room homes. Initially there were 334 farms with 20 acres, 64 with 30 acres, and 102 containing 40 acres. Eichenbaum and Williams located house sites on each of the four corners of a section, and roads adjoined every side of square plots. Work on the first three houses began in July 1934. By August,

---

[44] Wade, "Founding of Dyess Colony."
[45] Streissguth, 10.

construction workers had four houses ready for occupancy and 30 sites cleared for construction. An August 19, 1934, *Arkansas Gazette* described houses being built. "The cottages are models of economical farm house architecture. They are built to meet the needs of a farm family. Their construction is simple but durable. They will be sturdy, comfortable residences of dignified appearance." House No. 1 became sleeping quarters for administrative personnel. Carpenters used No. 2. During that first year workers completed 146 farm houses and 20 town residences. Substantial results came about because supervisors used teams for identical work and enforced strict schedules. A seven-man crew laid three foundations per day. Ten men constructed five-room house frames in 16 hours. Four workers shingled a house in six hours. Eight carpenters put up siding in 16 hours. Eight men completed floors and ceilings in 16 hours. Three workers installed windows, doors, and hardware in 16 hours. A seven-man crew built barns in six hours. Six painters primed a house in five hours and completed painting in six hours. This efficiency extended throughout the project's development.

Above, by the time the first families arrived in Dyess in 1934, roads had been cut through the colony and ditches dug to assist with drainage, but conditions remained primitive at best. Below, a Dyess Colony farmstead showing house, barn, chicken coop, and early planted field. **National Archives, Records of the Work Projects Administration**

6
Bound for
The Promised Land

Prospects to occupy Dyess farmsteads came from every Arkansas county. They heard about this opportunity from several sources, including county agents and rural rehabilitation program participants. Notices placed in newspapers and on radio alerted people to the project, and they in turn shared information with neighbors. Data about applicants came from written applications, references, and investigations. Using these sources FERA gained personal information about a prospect's occupational experience, education levels of family members, and why a wife and husband desired resettlement. The agency also wanted to know if applicants had previous experience farming. Additional personal information sought during this process included age-sex composition of families, general health, and stability. References came from bankers, teachers, medical authorities, ministers, community leaders, and government officials. These sources indicated an applicant's credit history, capital resources, intelligence, health, religiosity, and cooperative spirit. In cases where data proved inadequate for a decision, social service case workers interviewed prospects. The first 13 families moved into their new homes in October after passing various levels of examination and being accepted by Arkansas Department of Social Services. Author Dan Pittman believes that the "early success of the Dyess Colony lay in the careful screening of applicants." Interviewers accepted only experienced farmers who "through no fault of their own, had lost their farms in the economic depression."[46]

Despite robust efforts to recruit colonists, some officials expressed reservations about the speed of their process. Amy Pryer Tapping, a federal social work supervisor, prepared a report for Arkansas FERA about Dyess recruitment in March 1935. She noted that 104 application files remained in the state office and pointed out a major concern. "In your list of applicants for Dyess Colony there evidently have been a good many rejections. I am wondering if some further check on the cases before they come for final approval by the state office would not make for fewer rejections." Tapping recommended that field representatives review files and talk to case workers who did applicant interviews to ensure sound selection decisions.[47] Referenced files had been rejected for many reasons, most notably because of incomplete case histories. They typically lacked a sufficient description of family members and enough employment history, particularly with respect to farming. Comments by references seemed vague. Inconsistencies occurred between narrative information and case worker analysis. The most common reasons for rejections turned out to be mental attitude; extreme political views; social habits; excess family size for available houses, and health conditions.

Another status report in April 1935 also indicated insufficient progress in processing applications. It listed 46 houses ready for occupancy, but only six families authorized to move in. This report itemized the following: "[Files] in hands of committee, 17; ready to be submitted to committee, 5; in hands of special investigator, 48; pending completion of information, 19; rejections to be reconsidered, 18; total under consideration, 107." Though persons involved in selecting colonists claimed to be applying a rigid review process, internal reports and correspondence suggested that urgency became a priority.

---

[46] Dan W. Pittman, "The Founding of Dyess Colony," *Arkansas Historical Quarterly*, 29: 4 (Winter 1970), 320.
[47] Amy Pryer Tapping, "Report on Colonization Project No. 1," Emergency Relief Administration Social Service Division, March 25, 1935. Everett Henson Collection, Dyess Colony Archives #2016-33-091, Arkansas State University.

Administrators stressed that efforts be made to "speed up the selection and moving of families to the colony." One goal called for selecting "prospects both from relief rolls and RR" in order to place in the "colony by October or November 1, 1935, those who have been accepted and approved through the summer months."[48]

In spring 1935, RA restructured FERA's departments of subsistence homesteads and rural rehabilitation, but this Arkansas project remained a FERA responsibility. Tugwell of RA and Westbrook's FERA agreed to have management sections of each agency work jointly to select colonists. A social services division of Arkansas FERA would assist by providing county case records and applicants from relief rolls in Arkansas counties. FERA agreed to provide funds for maintenance of selected families until the project moved to RA for management. The magnitude of state employment problems became clear when Dyess spoke to Arkansas FERA members at a July 8, 1935, conference. "We have 35,000 men and women now on work projects of the Relief Administration, and we must either get these projects approved as federal projects or get new projects in their place and absorb the works into the Works Progress Administration. In addition, we have 20,000 other relief clients not on the work program. If they are able-bodied, we must develop additional projects to absorb them, and if they are not able-bodied, we must forget all about them and tell the State of Arkansas that they are the responsibility of the newly created State Welfare Department. That is the picture."[49]

As of August 1935 about 277 farm houses had been completed, with 16 under construction. These houses must have seemed like palaces to poor families who moved into them. They had modern kitchens, tongue and groove floors, glass panes in the windows, and several closets. Water came from individual wells and hand pumps. Separate septic tanks stored sewage. When completed, houses with

---

[48] Arkansas Department of Social Services Report on Colonization Project No. 1, April 17, 1935. Everett Henson Collection, Dyess Colony Archives #2016-33-024, Arkansas State University.

[49] Background Section, "Final Report and Physical Accomplishments of the Arkansas Work Projects Administration," March 1, 1943. Everett Henson Collection, Dyess Colony Archives #2016-33-031, Arkansas State University.

three, four, and five rooms were projected to cost $1,007, $1,175, and $1,379 respectively. However, purchase prices remained open since final costs would include some pro rata charges for roads, ditches, and other improvements. Additional structures that Eichenbaum designed included buildings at the colony center. In the middle of a 160-acre area he located three brick buildings placed in a semicircle. The middle one, a stately building, contained administrative offices. One building housed the commissary and the other a café and shops.

Applicants fortunate enough to be chosen for a Dyess inspection experienced a highly orchestrated performance by colony management. After guests finished breakfast at 8 a. m., guides gathered their charges. During tours a selection specialist questioned and observed prospects carefully to form an opinion as to whether the family would be a suitable addition. As one might expect, men generally paid more attention to the soil and farming opportunities. Some declined to enter available houses, leaving that choice up to their wives. In order to form a decision about living conditions families viewed a model house with all the latest furnishings. It included a day bed, a dining table with extension, four chairs, and a sideboard. Metal beds featured stained wood headboards. Most homes had stained panel walls, while some kitchens and bathrooms were painted a cream color. If a family brought its own furniture those items were integrated into their new house with help from a social service supervisor. During tours, guides questioned visitors about their commitment to farm life at the colony, and those with reservations received advice to withdraw their applications. After acceptable families selected a house, passed medical examinations, and completed interviews, Dyess gained new residents.

People who survived this application process and pulled into Dyess found two things for sure—a decent place to live and plenty of hard work to go with it. A. J. Henson recalled that "I was four years old when we moved to Dyess. I do remember moving into a nice clean house that had windows and everything," but no electricity. "We had an outhouse, a barn and a chicken house. We lived five miles from Dyess [center]." They also found friendly neighbors who were "helpful to us." Henson said that he worked with his father in cotton fields "when you couldn't see the end of the cotton row. We always had to go to the field. When we got big enough to carry water, you could be water boy. When you got big enough to pull a sack or swing a hoe you were a full hand. Now I look back on those as being good times. I doubt for sure dad did."[50]

After a two-day journey in a truck the Cash family pulled into Dyess in March 1935. Ray Cash, father of Johnny Cash, heard on the radio that "we could buy 20 acres without any money down, and a house and a barn, and they would give us a mule, a cow, and furnish groceries through the year until we had a crop and could pay it back."[51] Since the Cash family had been on relief since 1934, Dyess looked like the promised land. Following approval for a place there they loaded their possessions in a truck and headed for the colony. In a July 7, 2000, *Atlantic Journal* article Johnny recalled singing, "I am bound for the promised land," as he rode in the back of a truck hauling the family to Dyess. They arrived at a house with two bedrooms, a living room, kitchen, dining room, and bathroom, along with a smokehouse and barn. Johnny recalls in his autobiography the muddy road to their new house and that the truck could not gain traction, so his father carried him the last 100 yards. Acreage surrounding their house resembled a jungle. "I mean a real jungle," he said. Trees, vines, and bushes grew tangled up in dense thickets, some of it under water. That first night the family slept on the floor amid empty paint cans, and the next morning they went to work. In various interviews Joanne Cash, Johnny's sister, described life at Dyess. "Our family life in Arkansas on the cotton farm was very hard work. My daddy was a very hardworking southern cotton farmer, and all he knew in his life was hard work. He was raised from what we call the 'old school,' where you literally had to dig life out of the dirt. Johnny had a different plan." Joanne said that when family members came home at noon to eat and rest they listened to a radio program called *Eddie Hill's High Noon Round Up*. "I remember one particular day the Carter family was singing, and

---

[50] A. J. Henson, Oral History Interview with Lisa Perry, May 19, 2007. Dyess Colony Archives #2018-10-005, Arkansas State University.
[51] Streissguth, 8.

I remember Johnny saying, 'Listen to the Carter family. One day you're going to hear me.'"[52]

Though the Henson and Cash families, along with other white applicants, found homes at Dyess, the colony denied black folks this opportunity. A government report addressing this issue has a paragraph that displays rank hypocrisy. "In none of our projects are we sanctioning any discrimination in regard to nationality, race, or creed. However, in localities where there exists deep-rooted prejudices they cannot be ignored. It is not our function to attempt to reform in a day age-old attitudes of the people in the localities from which we are drawing our families, though we cannot dodge the responsibility of following enlightened social practices. We need to give consideration to homogeneity within the group and harmony both within the group and between the group and the outside neighborhood."[53] In the outside neighborhood, meaning small towns and farms surrounding the colony, an integrated STFU gained members and energy in its fight against abusive landlords. As a result of the union's success merging blacks and whites with shared goals, a cynic might question the basis for this federal discrimination.

---

[52] Interviews with Joanne Cash: Aug. 11, 2011, Arkansas State University Radio-Television production students, Dyess Colony Archives #2018-11-001; Sept. 4, 2011, *Arkansas Democrat-Gazette*.

[53] John P. Holt, "An Analysis of Methods and Criteria Used in Selecting Families for Colonization Projects," Social Research Report No. 1, Farm Security Administration and the Bureau of Agricultural Economics, Washington, D. C., 1937, 7. Record Group 96, Farmers Home Administration, National Archives. Copy in Dyess Colony Archives # 2016-33-004, Arkansas State University.

Above and below, new colonists arriving and unloading belongings at their new home.
National Archives, Records of the Work Projects Administration

Above, WPA workers install privy for newly arrived colonists. National Archives, Records of the Work Projects Administration *Below, twins Josephine and Maxine Mitchell with their mother, Mary Mitchell, at an entrance to Dyess Colony.* Courtesy of Joie Mitchell Ward Clifton

7
How the Cow Ate the Cabbage

Colonists arriving at Dyess told interesting stories about their journey and experiences settling into a new way of life. Homer Joe Johnson, his father, mother, and three siblings moved to Dyess after sharecropping in Drew County, Arkansas. "We heard about Dyess in June 1934. We were sharecroppers." Typical of that time and place, his father had a sixth-grade education and his mother finished the eighth grade. The family rode to the colony in an eight-wheel wagon pulled by four mules, and it took two days and one night. "We spent the night of travel in DeValls Bluff [Arkansas], walking part of the time and riding part of the time. The team driver rode the mule on the left, hooked next to the wagon. We had to pay a toll charge to cross the St. Francis River near Forrest City. The driver had permits for this charge. We arrived with the bare minimum of housekeeping furniture, a milk cow, and a few chickens in a coop tied on the back of the wagon." Upon arrival they moved into a four-room house with a two-story barn and medium-size chicken house. The house had a front porch and small back porch. We were located one mile directly west of old center where the road turned south to the new center, on the left or south side of the road with the ditch cutting off about two acres of the 40-acre plot."

After settling into their home on Road 2 the Johnsons became members of a welcoming committee that met and befriended new colonists. Their relationship with outsiders proved to be less cordial. According to Homer, "Outsiders viewed us as second class, partly illiterate . . . and turned up their noses at us." However, "they soon learned how the cow ate the cabbage . . . and in no uncertain way."[54]

The Mitchell family had difficulty landing a spot at Dyess. A daughter, Maxine Clifton, explained that before acceptance they lived in Stone County near Marcella, Arkansas. A friend of the family told her parents about Dyess. "My parents, my twin sister, and I made the inspection trip [with] Mr. and Mrs. Lafayette Wilson and Mr. Wilson's mother, Grandma Johnson. The driver got lost and drove into Missouri, all the way to Poplar Bluff, before he got us to Dyess. At first our family was not approved for a Dyess place," but a friend intervened. "We received notice to have our household goods on the main road by 10:00 a. m. on a set day. When the time arrived, so did two trucks, a 1934 Ford and an International. It was late in the day by the time we arrived at Dyess. Daddy checked us in at the office and received a purchase order for groceries. After picking up the groceries it was dark by the time we arrived at our new home. The first morning, mom noticed that the sun came up in the wrong direction. That was not the only difference between our new home and our old one."[55]

Marlin Joe Roberts recalled his family's move to Dyess and what they found upon arrival. "Our parcel of land and home was House No. 541 on Road No. 13. I remember a large truck with a tarp over the bed hauling our furnishings to the intersection of roads 2 and 13. Our furnishings were unloaded onto an eight-wheel log wagon pulled by a small track caterpillar tractor which moved us up the muddy road approximately three-quarters of a mile to our new home. The house was white, which you could hardly see from the road, a distance of 75 feet, for the thickness of the woods between the house and the road. We lived at this location for one year."[56] Two relocations

---

[54] Homer Joe Johnson, *The Delta Historical Review*, Mississippi County Genealogical Society, Summer 1990, II:1, 15-17.

[55] Maxine Mitchell Clifton, "'Thomas Green' in collection of memories of the Dyess Colony," *The Delta Historical Review*, 2:1 (Summer 1990), 20-21. Dyess Colony Archives #2018-14-004, Arkansas State University

[56] Marlin Joe Roberts, Personal History prepared July 11, 2005 for Dyess reunion. Dyess Colony Archives #2018-10-003, Arkansas State University.

put them on a 40-acre farm about a mile from the community center until they moved to California in the 1950s.

Transporting families to the colony from rural areas throughout Arkansas sometimes turned into an ordeal. One trip report recounted several difficulties. "Everything worked lovely until they [family members] were loaded and started on the trip. About two miles of the five of rutty clay soil county road had been covered when it started a heavy rain. In less than 200 yards farther the truck wheels slipped off the ridges into the deep ruts, and they were permanently stuck. This happened before noon. The family had five small children, one of them a two-month-old infant, and it was on a cold, rainy, early March day. The driver could not locate a tractor in the neighborhood; the loaded truck was too heavy and stuck too deep for a team [of mules] to get them out." Fortunately the driver found a nearby telephone and called the Dyess transportation director for help. "The director left town in a small pickup truck to go the 30-odd miles to his assistance. He located a tractor, which had to go around through a field to get to the mired truck. On the way out to the highway, the pickup truck stuck three times, making it necessary each time that the tractor be detached from the loaded truck, go around in front and pull the light truck out, then go back to the load again. It was 2 a. m. before they finally got the load to the highway and started on to where they could find quarters to spend the rest of the night—and the mother with a two-month old babe in the cab of the truck."[57]

All sorts of difficulties cropped up during these trips. One family to be moved owned a wagon, several farm implements, 57 bales of hay, and a truckload of personal belongings. Unaware of the large number of items to be hauled, Dyess transportation officials had to make several trips to pick up their property. Another family possessed five truckloads of items. Two issues precipitated many transporting

---

[57] Eula Gallagher, "Resume of Procedure and Problems of Planning Inspection Trips and Moving Families to Dyess Colony," Family Selection Section, Resettlement Administration, April 25, 1936, 6. Record Group 96, Farmers Home Administration, AK-80, National Archives. Copy in Dyess Colony Archives, Arkansas State University.

problems: families to be moved who lived on poorly maintained dirt roads and incomplete lists of assets to be transported. Settlers listed items to be moved on a form designated RAMA 21. But this document did not allow for an itemized accounting of household goods. Nor did it ask for livestock, feed, and tools. As a result, many items went unlisted or understated. One bizarre incident occurred when a ferry boat carrying colonists and their belongings attempted to cross the swift Current River. A front slip dropped in the middle of the river, and the vessel took on water, causing it to sink with a loaded truck on board. Water came up to the hood, and workers attempting to salvage the truck needed two days and nights to recover the vehicle and its contents before sending them to Dyess.

Ed Wooten shared his experiences relocating to Dyess. Prior to the move his family lived in southern Arkansas so their father could work in the timber industry. "During the Depression it was 18 of us in that house we lived in. Daddy could always manage to have work and food. Back then, that's just the way it was during the Depression." Unfortunately for the family his father could not land a steady a job in the timber fields. To help out, Ed and his brothers "took care of horses and mules. We had about a hundred head that belonged to the government, and me and my brother, while our daddy worked what little work he could get, fed those horses and mules, which was a pretty good job." After reaching the colony "they loaded our house furniture in a log wagon with a D-6 CAT [tractor] pulling it down to Road No. 9." When they began to unload their personal items his mother expressed buyer's remorse. "My mother didn't like it at all, because the only thing cleared up here then was a few acres around the house site and the barn. Mama kept telling daddy not to unload that furniture, and daddy just kept unloading it." Despite their mother's misgivings, the Wootens moved into a two-bedroom house with a kitchen, living room, and a dining room. "We filled it up because there were three boys and one girl. We had to use that dining room for another bedroom." [58]

---

[58] Ed and Janie Wooten, Oral History Interview with Lisa Perry, May 5, 2007. Dyess Colony Archives #2018-10-001, Arkansas State University.

New colonists generally found areas around their house unimproved, requiring enormous effort to clear the land. Vera Clements recalled a common danger they faced—numerous poisonous snakes. "We were cutting trees and a friend of Daddy's was helping. All of a sudden me and Dad stopped sawing. Me and him always sawed the trees down. He said, 'Be still, sis.' He always called me sis when [we] were talking. He said, 'Stand still, sis, don't move!'" Her father told his friend to go to the house and get a gun. "I couldn't imagine what he was going to do with a gun and me standing still. There was a big rattlesnake right beside where I was standing. Dad shot it. It had rattlers on the end of it. For a long time Dad had them rattlers in the old guitar we had."[59]

---

[59] Vera Knight Clements, Oral History Interview with Emmett Powers, February 20, 2013. Dyess Colony Archives #2018-10-014, Arkansas State University.

Above, Hazel Kimbrough tending her garden. Below, Hershal Henson working in the field with two of his sons, A. J. and Coy. Everett Henson Collection, Dyess Colony Archives, Arkansas State University.

8
LIFE WITHOUT SANTA CLAUS

Soon after colonists arrived they began preparing land for crops. Administrators required that one male family member work as a day laborer on construction projects. The remaining men and older boys cleared, planted, plowed, and harvested their fields with one or two mules. Women and older girls generally did almost everything else. They tended to the young children, cooked, and kept up the household. They usually fed the hens and collected eggs. Women washed and mended clothes to keep family members clean and presentable. They tended the garden, which often had spinach, mustard greens, turnips, radishes, tomatoes, and onions. In addition to these chores, sometimes women went to the field and worked beside men.

Like all new arrivals the Johnson family quickly learned the practical nature of life in Dyess. They received coupon books with values determined by the number of family members and their ages. Colony managers gave them garden seed to start growing some food to eat. "We did just that, too. This was of great help to each family. We were issued a team of mules with harness and tools to farm the land. Some feed was issued to start with until we started growing some food for the animals and ourselves." The family's diet "sometimes was beans and potatoes to eat, but we had plenty of them. No one left the table hungry either." Gravy and biscuits for breakfast were measured, "two or three biscuits depending on how many it took to hem up the gravy in the plate."[60]

Colony officials charged to capital advance accounts the cost of livestock, implements, and supplies, which could total no more than $600 per year. Loans typically carried a five percent interest rate. The Johnsons and others used wages from day labor to supplement their income. Willing men performed work on colony projects, such as clearing land, and earned $1.60 per day. But colonists had to limit these earnings to $24 per month. Each male family member 18 or older might hire out for day labor as long as it did not cause a crop to be neglected.

Frances Wallace's family lived in Knobel, Arkansas, when her father "bought his place at Dyess, sight unseen." The day her family moved to the colony "snow was on the ground and cold, oh it was cold, but we had a place we called our own. That summer I chopped cotton. I could chop two rows of cotton in the morning and two in the afternoon because that's how grassy it was. So my Dad told me I didn't know how to chop cotton. He was going out there the next morning and show me how. But he chopped two rows, and I chopped two rows. We didn't go back after dinner, but he did rake up so we could go down and get most of the grass." Wallace always got her row clean "because Daddy told us we had better chop it clean, and I believed him. I'd be out there chopping sometimes by myself, and a cloud would come up. I'd say, 'Come on, rain,' whether we needed it or not, just so I could get out of the field." Wallace had only one sibling, a brother, but "we did it all ourselves. When we first moved there every bit of our farm was grown up. We cleared it. That's why I could only chop two rows in the morning that first summer."[61]

Wallace said that "One summer Daddy bought an old pull-top hay baler, and it took four of us to run that hay baler. One to drive the tractor; one to stand up front of the hay baler and make sure it didn't stop up; one to punch wires and block, and one to tie off. I punched the

---

[60] Johnson.
[61] Frances Forrester Wallace, Oral History Interview with Lisa Perry, May 19, 2007. Dyess Colony Archives #2018-10-002, Arkansas State University.

wires and blocked. I looked like a grease monkey every day when I went home, but we thrashed hay all over Dyess. We made quite a bit of money that summer making hay." After a day of chopping cotton, when Wallace had to help her mother clear the garden of grass and weeds, "I cursed underneath my breath all the time." In addition to cotton chopping Wallace disliked milking the cow. "I never learned to milk. I'd go out there and get maybe a half a pint or pint, and Mother would come out behind me and get a gallon. So I had the job of churning until Daddy finally got an electric churn." Her life there made Wallace industrious and frugal. "I'm not afraid of work, and I know the value of a nickel. I started to say a dollar, but I know the value of a nickel even. Neither one goes that far nowadays."[62]

Bernice Alma Burkhart described how their universe revolved around chopping and picking cotton. The school schedule depended on a picking season so children could help harvest crops. "They started earlier, no recess, etc. Kids would get home about 2 p.m. We paid our children for their picking, to teach them how to manage money. They did a good job. They most always bought their clothes. We'd of course help with costs, the big things, and they would always have their spending money. Allie [her husband] and I could never decide just what Travis [their son] did with his spending money." They knew he went to the picture show and supposed he spent the rest, eating and drinking. The girls would buy "makeup, something for their hair, and candy, gum & etc. They might have some change, but not Travis. Then he'd always manage to get them to share their candy with him." [63]

When crops were laid by, families gathered together. Women cooked, and men sat on porches and shot the breeze or competed in a game similar to horseshoes. Kids played tag or merry-go-round in the yard. During warm weather many youths often went swimming in the Tyronza River at a spot called the "blue hole." The river also allowed Helen Wright to go frog gigging. She said that "I ate the frogs that they caught all the time, and I loved them, but I wanted to go gigging

---

[62] Wallace.
[63] Bernice Alma Burkhart, unpublished and undated memoir. Dyess Colony Archives #2018-13-001, Arkansas State University.

myself." So her father took Helen gigging one night. "Well he got me in the boat, and we both got on [carbide] head lights. I'm in the front of the boat and he's in the back. We're coming down that river and I'm so happy, I'm gigging them frogs." That night she broke into a sport generally reserved for men and boys.[64]

A slack season also allowed time for fishing in the Tyronza, wiener roasts, cake walks, and dances. Music usually poured out of a Victrola at parties, but occasionally musicians played songs on guitars, banjos, and fiddles. Families watched picture shows at the community theater, and teenagers gathered at the "pop shop" to mingle and flirt. When farmers had a little money in their pocket after harvest, entertainers came to towns near Dyess. They typically stayed for a week, attracting people from nearby communities. Fair operators set up rides and tents for vaudeville acts, drawing large crowds. Ed Hardin remembered them well. "Shows would come to Dell [a town near Dyess] in the fall of the year. Tent shows would come in. They put up a big tent and put up seats under there. I got to go to several of them because my granddaddy was the mayor, and he gave me a penny. I was fortunate."[65]

Colonists also showed up to support their school's athletic teams, basketball and football. Everett Henson played football and mentioned that a Dyess team set an unusual record. "Before I started playing, a few years before that, they had only six playing football here in Dyess." But the team won a state championship in six-man football. Colony athletes gave folks something to cheer about. Jean Ann Jennings remembered the school song. "We are loyal to you Dyess High. To our colors we're true Dyess High. We will back you to stand against the best in the land, because [we] think you are grand, Dyess High we're true Dyess High." Though generally from poor families,

---

[64] Helen Johnson Wright, Oral History Interview, Memories of a Lifetime Project Team, October 20, 2017. Dyess Colony Archives #2018-12-007, Arkansas State University.

[65] Ed Hardin, "Fun Things to Do in the Early Years of Mississippi County," compiled by Jean Ann Cannon Jennings, posted on Dyess Colony private website. Copy in Everett Henson Collection, Dyess Colony Archives #2016-33-288, Arkansas State University.

colony athletes took pride in their appearance. Jennings stressed "How proud we all were when our boys walked into the gym in their suits and ties. What a [sight] they were."[66]

Dyess residents spent much of their leisure time in church. When the colony's first families arrived, various denominations worshipped together in a livestock barn at a construction camp until completion of the community center. Congregations sat on logs and attempted to drown out with hymns the hee-haws of mules. Despite primitive facilities they created a community church organization governed by a nine-member board. Three stewards supervised pulpit activities, three deacons handled general duties, and three trustees took responsibility for property. The Presbyterian Board of Home Missions furnished a pastor with a non-denominational spirit, hymn books, and other materials. In November 1935 a survey revealed the following memberships, though categories may be imprecise: Methodists, 67; Baptists, 87; Presbyterians, 2; Pentecostal [most likely undercounted] 4; Church of Christ, 4; Missionary Baptists, 4; Christian [probably Disciples of Christ], 7; Holiness, 32; Catholic, 2; Apostolic, 3, and Assembly of God, 2.[67]

Separation of different denominations into homes of members soon brought segregated congregations, which eventually relocated to newly constructed facilities. Various congregations got along well for the most part, but in some instances friction developed. A colony official described one example of tension generated by "some Holy Rollers, especially among the more ignorant." Managers expelled a female "Holy Roller" from the colony because she "kept the colony more or less stirred up by her religious outbursts."[68] Two researchers

---

[66] Jean Ann Cannon Jennings, Oral History Interview, Memories of a Lifetime Project Team, October 20, 2017. Dyess Colony Archives #2018-12-002, Arkansas State University.

[67] *Dyess Colony, Inc. Project Book*, Chapter III. Community Operation and Management, Section F. Education, Recreation and Health, Religious Groups. Records Management Division, Resettlement Administration. Record Group 96, Farmers Home Administration, AK-80, 700, National Archives. Copy in Dyess Colony Archives #2016-33-032, Arkansas State University.

[68] Ibid., Chapter III, Section B. Family Selection and Census-General.

studied the characteristics of seven resettlement colonies with respect to religious practices and interviewed 415 of 484 families at Dyess for their survey. They found that "by far the most important institution, as would be expected in rural areas, was the church, and formation of some type of worship service had demanded priority over all other types of social agencies."[69] But some colonists interviewed about their religiosity may not have been entirely forthright. According to Ann Blue, church deacons "played poker over on the Tyronza River bank trail—the trail going to naked hole." She also revealed one of Johnny Cash's odd religious practices. On Sunday mornings "J. R. would climb the 150-foot water tank and study his Sunday school lesson," and perhaps other things as well. "The big boys told me some of the Dyess girls skinny-dipped in the tank."[70]

During the colony's first months, days took on a country normalcy. No couples divorced. Six deaths occurred, but doctors traced them to conditions unrelated to relocation. Three weddings and 14 births indicated that life was good. However, for many farmers who made it to Dyess, staying was another matter. Administrators observed men and women carefully for several months to decide if they exhibited the requisite industriousness and personal habits required of residents. Colonists underwent a one-year trial period before receiving final acceptance. A farm manager and a home economist made this determination, and not all settlers made the cut. Administrator comments indicated that some colonists held the mistaken belief that a place in the colony constituted a gift rather than an opportunity to work for a new start in life. Management encouraged those persons to move out.

Dewey and Elsie Cox, who moved to the colony in 1936 from Pike County, were among the many colonists who seized the opportunity and worked hard to make a go of it. But the youngest of their nine children, Larry Cox, agreed that it was not the answer for everyone. "Some people did fail, absolutely they did," he said. "But

---

[69] Charles P. Loomis and Dwight M. Davidson, Jr., "Social Agencies in the Planned Communities," *Sociometry* II: 33 (July 1939), 28.

[70] Blue.

for the majority of the people that understood what it was about and committed themselves to the work, it was a great success."[71]

In a September 22, 1935, *Arkansas Gazette* the Dyess home economist, Fern Salyers, expressed frustration with the attitude of a few colonists. "Some of them came here under the mistaken impression that they would find Santa Claus." Another administrator, H. Spicer, criticized farmers who did not adjust to gumbo soil. Transplanted from hill counties they "just couldn't become acclimated to the capricious buckshot soil of the colony. It can be plowed best under two conditions; when it's very dry or it's so wet it leaves water in the furrow behind the plow." Some colonists took his advice, but others did not. They waited until spring, "when plowing it was like trying to plow rubber." Spicer also complained about a farmer who could not get along with his mule. "Well, you've got to have more sense than a mule to be able to manage a mule."

---

[71] Larry Cox, Oral History Interview with Ruth Hawkins, July 7, 2018. Dyess Colony Archives #2018-33-003, Arkansas State University.

The Administration Building anchored the Dyess Community Center, which included public buildings to serve the colony and homes for teachers, medical personnel and federal workers. Above, Everett Henson Collection, Dyess Colony Archives, Arkansas State University; below, National Archives, Records of the Work Projects Administration.

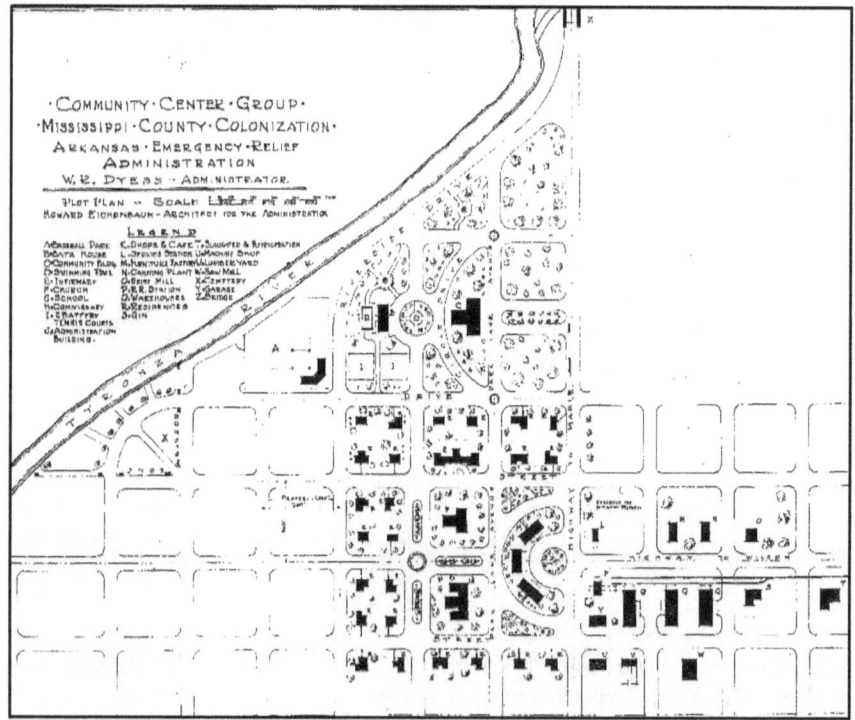

9
AN OPPORTUNITY WITHIN REACH

At the end of 1935, administrators surveyed colony inhabitants. A family selection and census report described what they learned. "We found in studying this group that the families had considerable farming experience. The 83 indicated as 'unskilled' should be listed as farmers with no other occupational training. I was particularly interested to know that the larger numbers of colonists were renters, closely followed by day laborers, with only a small number of sharecroppers. Of the 157 families sent to the colony 20 have been sent home, leaving 137 families composed of 756 individuals." One family from Clay County created dissension. "The man was not familiar with southern customs and race distinction meant nothing to him. He solicited boarders from the Negro workmen at the colony and took them into his home. He spent little time on his farm and made no progress whatever. He was sent back to Clay County."[72] One wonders if the man's forced departure had more to do with his choice of boarders than his farming practices.

This census made an interesting point about the relationship between age and success at Dyess. "Experience with the colonists seems to show that a young couple cannot undergo the hardships that the older people can. A man from 40 to 50 years of age with four or five children is the better type and the best one to work with for he thinks of his children and what he will leave for them. The young people feel that they can go back to their families when things go wrong and

---

[72] *Dyess Colony, Inc. Project Book*, Chapter III. Section B, Census-General.

consequently do not make the effort necessary for success." A Dyess social worker received a letter that conveyed the regret of two young colonists who returned to their previous home. "Sorry we left. Was just homesick I reckon. After we got here and saw everybody we was ready to come back up there. But we decided too late. I guess we ain't hungry or kneed [sic] nothing. We are just homesick. I sure do hate we left. I will never be satisfied no where else but up there."[73]

Analyzing the project's start-up finances can be a daunting task. Probably the most reliable accounting of the colony's early expenditures and balance sheet values came from an audit prepared by certified public accountants at Russell Brown & Company in Little Rock. Based on financial statements dated February 29, 1936, they recorded entries from February 1935 to the February 29 records. "The colony received cash grants from the Emergency Relief Administration amounting to $2,306,250.00 and on February 29, 1936, owed $53,703.56, and had a capital surplus of $174,034.97, a total accountability of $2,533,988.53. With this amount the colony has purchased approximately 16,000 acres of land at a cost of $136,994.48; constructed 500 farm houses at a cost of $942,417.49, including out houses; constructed 38 residences at the community center at a cost of $65,440.72; constructed permanent buildings at the community center at a cost of $190,692.09, including the administration building, community building, café building, store building, gin, and other buildings; purchased machinery and equipment, [with] a depreciated value at February 29, 1936, of $111,932.12; built a network of roads throughout the colony at a cost of $399,980.39; built road bridges at a cost of $51,347.89; drained the entire colony at a cost of $84,595.22 for drainage ditches; built a railroad spur track from Hitt, Arkansas, to the colony at a cost of $35,908.49; constructed streets, curbs, gutters ,and sidewalks at the community center costing $12,601.01; a water system at the community center costing $26,549.55; sewer system costing $6,530.38, and power lines costing $11,113.44; had cash and receivables

---

[73] Ibid., General.

on hand amounting to $224,647.17, furnishings amounting to $220,397.45, and other assets amounting to $10,840.64."[74]

Confirmation of difficulties associated with keeping tabs on colony finances came in a February 1936 memorandum from WPA. In it Sharp announced to residents that assigning prices of homes would be delayed until completion of most building construction and calculation of costs. He announced that Dyess Colony Incorporated (DCI) would assume legal title to the entire project and arrange sales of home sites to colonists. "We want you to know that our every intention has always been to make available to you a home at a fair price and payments extended over a period of years. We are planning our program in such a manner that there will be no hunger or actual want in the colony." Sharp closed his message with assurances that administrators "will give you every cooperation [and look] forward to the day when you and your family can be independent in your own home."[75]

Cost overruns provided ammunition for critics sniping at this resettlement community. Negative reaction of many conservative legislators to such programs came out in a March 31, 1936, *New York Times*. It reported that the Republican National Committee accused President Roosevelt of sponsoring farm communities "communistic in conception. Resettlement Administration is establishing communal farms which follow the Russian pattern." Agricultural power centers such as the American Farm Bureau Federation and many state extension service agents resisted federal projects that assisted impoverished farm families. They considered such programs to be threats to their control of labor. Extension agents were employed jointly by the federal agriculture department, a college of agriculture, and the county they serviced. Assigned to convey valuable information to

---

[74] Russell Brown & Company, "Report to the Office and Directors, Dyess Colony, Incorporated," Little Rock, Ark., Feb. 29, 1936. Everett Henson Collection, Dyess Colony Archives #2016-33-071, Arkansas State University.

[75] Floyd Sharp, "Memorandum to All Residents of the Dyess Colony," Feb. 10, 1936. Works Progress Administration, Little Rock, Ark., WPA-674. In *Dyess Colony, Inc. Project Book*, Exhibit E, Chapter III, Section G. Terms of Occupancy.

farmers, extension agents answered to county officials, which often included prominent landowners. Planters and in some cases extension agents complained that RA directors did not "share their objectives and ideals [and did not] harmonize with the spirit of the people." Dan T. Gray, director of Arkansas extension services at the University of Arkansas, agreed. "The Resettlement Administration is doing some things in Arkansas and teaching some things that the people connected with the college of agriculture cannot endorse."[76]

Despite criticism by powerful business interests and their allies Dyess became a coveted land of opportunity for many needy Arkansans. An April 30, 1936, Resettlement Administration (RA) letter to Dyess officials explained the status of colonist applications. At that point the colony had 480 approved applications with 13 families being moved and 10 soon to be transported. Five applications were pending, with one accepted and the RA waiting for a contract. In one case a wife did not accompany her husband on an inspection, and her medical statement did not appear satisfactory. In another case a young couple was anxious to join the colony. The wife was a daughter of colonists from Sebastian County, and they had been referred by RA. Colony managers considered them a poor risk, however, but intended to make further investigations.[77] Another family would be accepted if the man's health improved. He became sick before starting an inspection, and colony doctors were treating him for pneumonia.

Even when included, certificates of good health often contained ambiguous information, and Dyess administrators placed great emphasis on the health of colony residents. Preventive medicine became a valuable asset at Dyess. Poor Arkansans had little access to medical treatment and usually no money to pay for it. Not so at Dyess. Lulu Turner, a public health nurse, each year carried on campaigns to protect children from contagious diseases, particularly typhoid and diphtheria. She considered typhoid a dreadful killer, particularly

---

[76] Baldwin, "The Resettlement Administration," *Poverty and Politics*, 116.

[77] National Archives, Records of the Work Projects Administration. Copy in Dyess Colony Archives #2018-15-004, Arkansas State University.

between June and October. Even when victims survived, the disease left them with weakened constitutions and more susceptible to other medical problems. Turner recommended inoculations every two or three years. The medical staff also stressed the importance of diphtheria vaccinations since many children were susceptible to that disease. Most diphtheria deaths occurred during the second and third years of life so doctors urged colony parents to have their children immunized. They also fought an ongoing battle with malaria and advised residents to dispose of containers of water such as old tires to eliminate mosquito breeding pools.

Medical personnel saw a variety of problems, some real, some unreal. Ann Blue remembered her stays at the hospital. "I know that I had my tonsils removed when I was in the third grade. I remember very well that I was laying in that bed and I just felt like crying because I thought everyone thought I was going to have a baby because I was in the hospital. I thought that was the main reason you went to the hospital—was to have a baby. And I thought, Oh my gosh, they'll all think I'm having a baby. But then I was only like about nine years old, so I guess I can be excused for that. I think I was in the hospital a week when I had my appendix out."[78]

When folks at Dyess marked their formal dedication on May 22, 1936, many major improvements had been completed, including 500 farmhouses. A permanent administration building stood at the colony's center, and nearby a cannery, cotton gin, commissary, community building, post office, café, and hospital. Records indicated that 481 families called the colony home. They cultivated crops on approximately 4,875 acres. Unfortunately, William Dyess did not live to see their progress. He died in a plane crash that previous January, but his son Billy attended the ceremony and dedicated a fountain at the colony's center to honor his father. Notables at events included Futrell, Westbrook, Eichenbaum, and Sharp. About 4,000 people attended festivities that day and night. During the evening colonists staged performances that may have been difficult to forget. Among the acts—

---

[78] Blue.

Oscar Smith and his wife performed a song composed by Mrs. Smith. Jim Gardner did a violin solo. Rosnel Hinesley gave a reading titled "Back in Squashville," and Arthur Craig tap danced.

The *Colony Herald* reported those memorable events and others as well. The newspaper published bimonthly from spring 1936 to the end of 1937, when it fell victim to budget cuts. Though considered to be a local newspaper, the *Colony Herald* included national and international news. A January 8, 1937, edition ran stories about a General Motors strike in Detroit, the former King Edward of England, and George VI's proclamation as king. Still, the paper made plenty of room for local news, and its editors claimed that some residents paid postage so that copies could be mailed to relatives outside the colony. One edition proved that Johnny Cash was not the only writer in his family, though he fared better than others. A May 29, 1936, paper published a poem by Roy Cash, Johnny's oldest brother. Entitled *Wild Western Outlaw*, it described some adventures of Banty Bill and Kansas Jake, and its beginning was not promising:

> *"We found that mining dident [sic] pay,*
> *And looked for some soft stake.*
> *We picked the game of 'Frisco*
> *Frank,'*
> *And after slinging lots of lead,*
> *We took five thousand from his*
> *Bank*
> *And left two Hombre's [sic] dead."*

The poem continued for another 39 tortuous lines and suggested that the newspaper must have had a slow news day and many column inches to fill.

In addition to bad poetry Dyess had other peculiar amusements. A June 19, 1936, *Colony Herald* described several. "A program of fun and frolic was presented at the Community Building... when Homer Wooten and his group of players gave a number of comedy acts," including "Crazy Willie," "Baby Elephant," and the "Book Agent." Apparently they brought "gales of merriment" from the audience. The entertainment included Wooten's impersonation of a Dyess farmer entitled, "Everything is Funny Down on the Farm." This, despite the fact that during the Great Depression very little was funny down on the farm, even at Dyess. Entertainment continued when members of the Live Wire Club threw a "tacky party." Each woman prepared a box of refreshments, and men drew names to determine "eating partners." According to the newspaper, prizes were awarded for the tackiest costumes, and Mrs. Hanley and Mr. Caraway won them.

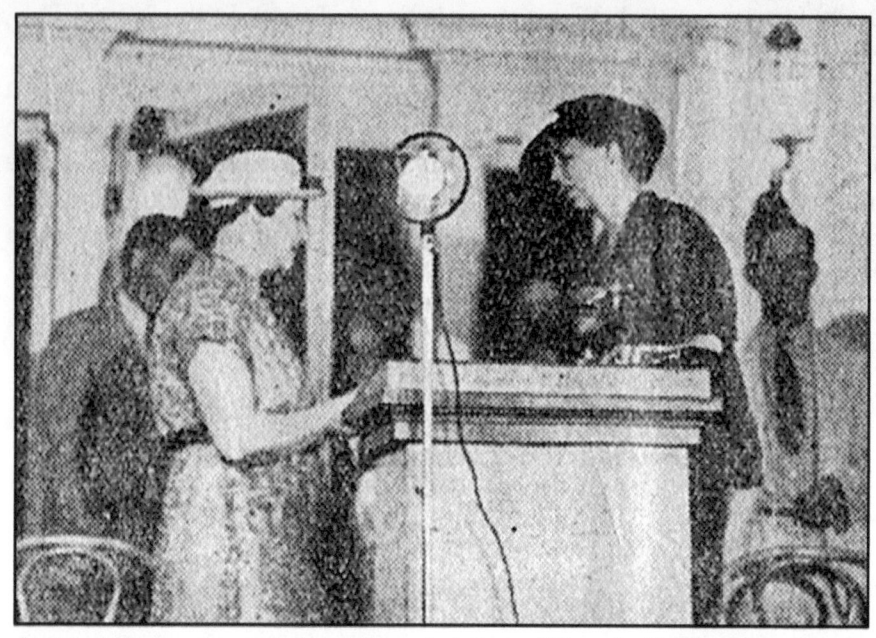

First Lady Eleanor Roosevelt, top right, spoke to Dyess Colony families from the porch of the Administration Building in 1936. After her visit she sent books for the Dyess library (below). **National Archives, Records of the Work Projects Administration.**

10
ACROSS A FRONTIER

Setting aside theatrical and literary abilities of colonists, Dyess had become a remarkable success in many ways. Two reporters, Jonathan Daniels and Phil Kinsley, put the community into perspective. In separate accounts they described their impressions while driving there. After observing many rundown Mississippi County tenant shacks along the way, Daniels emphasized a stark demarcation. "I crossed the colony line like a man moving across a frontier." During Kinsley's trip he drove through plantations and passed row after row of farm workers' cabins. Upon entering Dyess the reporter saw a remarkable transformation. "The unprepossessing homes of sharecroppers gave way to broad roads bordering neat homes with front yards flaming in purple feathery flowers that wave like banners of dignity and freedom. The houses are painted, and there are barns and out buildings, glimpses of corn and gardens, hay and beans, cows, pigs, and chickens."[79]

F. Arthur Currie designed the impressive appearances of Dyess homesites. The colony's landscape gardener, he advised homeowners about the importance of appearances. Currie stressed that improvement efforts "changed appearances we can all see as we pass along the roads where zinnias, poppies and numerous other flowers are in blossom making brave splashes. Let us work toward a definite plan so that the flowers, shrubs and trees which we set out and the care we give them will make for a pleasing appearance as a whole, giving

---

[79] Cannon Jennings, Posted June 23, 2005.

due consideration to a balanced beauty of line and texture."[80] Balanced beauty became especially important when Dyess participated in an annual National Better Homes Week. E. S. Dudley pressed families to gussy up their property to impress visitors during this celebration. However, his recommendation for a kitchen table seemed silly. In a May 1, 1936, colony newspaper Dudley described in detail how residents should determine the proper height of their kitchen table. A man should stand erect, he said, "feet flat on the floor and drop your arms down to your sides. Now hold your hands out in front of you, palms down as though you were placing them on a table. If the table is the proper height for you, your arms will not be bent at the elbow and you will not stand stooped over." Dudley then added, now "imagine how much . . . you will enjoy a meal." Intrusive advice came from all directions. In a memo to colony women, Fern Salyers, a home economics specialist, instructed them about personal hygiene. She stressed the following:

- Take a tub bath with soda and ammonia at least two times each week.

- Brush your teeth twice each day with soda and salt.

- Wash your hair twice each month with soft water, soda, and vinegar.

- Clean your stockings every day.

- Change your underwear every day.

- Keep top clothes clean and free from odor.[81]

Colonists made many improvements to prepare for a major celebration on June 9, 1936, when Eleanor Roosevelt came to have a

---

[80] Unidentified newspaper clipping in the Dyess Colony Archives, Arkansas State University.

[81] Everett Henson Collection, Dyess Colony Archives, Arkansas State University.

look. She arrived about dark that day for a tour led by Sharp. Given her late arrival the First Lady visited only one family at home that night. The Echols household welcomed her, and Bob Echols described the event. "She met all of us and was introduced to us. She was shown through our house and was interested in a lot of things, including my brother's stamp collection. She said that her husband liked to collect stamps. My sister played a piece for her on the piano. When she was ready to leave, she shook hands with each one of us kids again and called us all by the right name. Now how a person can remember names like that I don't know."[82] Later that evening Mrs. Roosevelt addressed a crowd of about 2,500 people at the administration building. A June 10, 1936, *Arkansas Gazette* reported her speech. "I don't suppose you realize the interest with which we in Washington have been watching your experiment here. We believe that the unfortunate people of this country would rather work out their own salvation than to accept charity." After her comments the First Lady shook hands with people in the receiving line and had dinner at the colony café. In her June 11 syndicated newspaper column Mrs. Roosevelt described her impressions of Dyess residents. "As I looked into their faces as they came by and at the children who slipped around and in and out, I decided that they had character and courage to make good when an opportunity seemed to be within their reach."

   The First Lady's interest in the colony had a personal connection that involved the Holland family. Born in 1924 at Siloam Springs, Arkansas, Bob Holland contracted infantile paralysis (polio) when six years old. He discovered the illness on a Sunday morning after his father Alva came into his room and told the boys to get up and get dressed for Sunday school. "My twin brother and I slept together. He jumped out of bed. I jumped out of bed and hit the floor. I didn't know what happened to my leg. I couldn't stand." His leg went limp from the hip down. A Kansas City doctor visiting the town did tests and made this dreaded diagnosis. Bob's mother, Geneva Holland, read about Franklin Roosevelt, at that time New York governor, having

---

[82] Mark Randall, "The Best Proposition a Poor Man Ever Had: The Founding of Dyess Colony." A research paper prepared for Wilkerson-Freeman (Spring 2010): History 6282, U. S. in Crisis 1929-1949, no page number.

polio, and she wrote Eleanor Roosevelt about Bob's illness. Geneva asked about treatment at the Warm Springs, Georgia, rehabilitation facility where Roosevelt went seeking a cure. Mrs. Roosevelt assisted with the boy's admission there, and he remained for nine months. During visits to Warm Springs Mrs. Roosevelt followed Bob's progress, and Governor Roosevelt took an interest as well. After the boy returned to Siloam Springs his three brothers and a sister hauled him to school in a red wagon, and doctors fitted him with a heavy steel brace.

Geneva continued to exchange letters with Mrs. Roosevelt, who encouraged the Hollands to seek acceptance at Dyess. Alva had lost his job and their house during the nation's economic collapse. Though he raised cattle and knew nothing about cotton production, which should have disqualified them, the family gained a place at the colony. This indicated that someone powerful may have influenced the selection process. Geneva's friendship with the First Lady continued, and crates of books from the White House arrived at the colony library where Mrs. Holland was librarian. With Mrs. Roosevelt's assistance and help from other benefactors, the facility amassed about 2,000 volumes and 1,400 members. It became a center of learning for colony children, with story hours, songs, and plays. The library also sponsored book review contests. Its importance in colony life confirmed that Dyess parents wanted more for their children than a sharecropper's existence.

The Hollands acquired a house, barn, mule, and implements at Dyess, but row-crop farming became a new experience for the family. They worked 20 and later 40 acres, and Bob called it "rough, rough work." His mother went to the field, so Bob stayed home and cooked. He admitted that "about the only thing that I ever did out on the farm was . . . sit up on the hay rake and rake hay." He said that despite hard work, living in Dyess was "one of the nicest, best memories I have . . . because everyone there was on the same plane. There were no rich people. They were broke when they got there, including us." Despite his many difficulties Bob remembered one scene that haunted him. While living in Siloam Springs the Hollands had a garden and enough to eat, but many of their neighbors did not. His father took Bob and his twin brother to "an old automobile dealership that had moved out, and they had four or five or six long tables and No. 3 washtubs with broken up crackers and four or five or six big jugs of soup." Their father told his sons, "I want you boys to see this." People were "going in there getting them something to eat. They'd give them a tin cup and dip the soup, and hand full of crackers." These hungry people would "stand up against the wall and eat it. I remember that, it made an impression on me."[83]

---

[83] Bob Holland, Oral History Interview with Emmett Powers. June 14, 2013. Dyess Colony Archives #2018-10-016, Arkansas State University.

The Dyess Colony hospital, above, and the commissary store, below, were among the enterprises expected to function as cooperatives, with operating costs and management coming from colonists. Everett Henson Collection, Dyess Colony Archives, Arkansas State University

11
PRINCIPLES OF COOPERATION

Dyess Colony Incorporated (DCI) became owner of project property in 1936 when Hopkins and Tugwell sold colony assets for one dollar to this corporation. However, they packed DCI's board of directors with WPA administrators in order to control federal money being spent at the community. Sharp, state director of WPA, became corporate president. To exert additional control Hopkins named Westbrook management counsel. DCI's board monitored all phases of colony business using a resident manager, E. S. Dudley. Despite this chain of command, Hopkins made clear in a letter to Sharp dated August 24, 1936, that he depended on Westbrook to implement policies for "administration of the community."[84] Thus colonists had Dudley looking over their shoulder. Dudley had Westbrook looking over his shoulder, and Westbrook had Hopkins looking over his shoulder. So with respect to project management there were a lot of people looking over a lot of shoulders. Long-term stability of settlers continued to concern Westbrook. He stressed that they should be prepared to endure an initial period during which satisfaction from farming was prospective, not immediate. Westbrook stressed that reliability had to be a cardinal virtue.

Despite the fact that Dyess became a desirable place to live, outsiders ridiculed life there. A critic from Pine Bluff, Arkansas, made several sarcastic observations before admitting the success of colony

---

[84] Harry Hopkins to Floyd Sharp, Aug. 24, 1936, National Archives, Record Group 96, Farmers Home Administration, AK-80. Copy in Dyess Colony Archives #2018-15-003, Arkansas State University.

residents. "Well, it's this way. The government spends a million dollars or so to buy a 40-acre farm for a down-and-out sharecropper. They give him a mule, a bathtub, and an electric shoelacer. They lay a railroad track to his house to carry the tons of forms he has to fill in. A bunch of experts figure out his milking I. Q. Lo and behold, they teach his wife how to hook rugs and can beef and spinach, and they show the feller how to plant soybeans and prune an orchard—and by darn, Luke, them government people can actually do it! After we poke fun at their red tape for a year or two they ups and proves their experiment is self-liquidating, [and] that the feller is making his payments and raising a family, too. And I don't know who's more surprised, me or the cropper."[85]

Westbrook did not share this wag's sense of humor. In August 1936 he addressed colonists at the community center building and shared his optimism. The *Colony Herald* reported his speech in its August 14 edition. His remarks deserve extensive consideration because they reveal details of his plans and philosophy with respect to the settlement. Westbrook emphasized that he intended to work with the colony's board and management. He then assured listeners that by owning their homes and achieving economic self sufficiency, "never again need you be assailed by the terrible fear that you and your family will have no roof over your heads, nor that stark hunger will overcome you. There is more real security here in the little homes that you live in than there is in the finest mansion in the most fashionable and arrogant residential section of Memphis."

Following some observations about colony recreational facilities, Westbrook got down to business, outlining matters critical for a farmer's success. He anticipated that families would need 70 to 80 percent of their overall production for personal use and colony sales. Thus 20 to 30 percent should be exported to outside markets either individually or collectively. "It is necessary that you have cash crops or

---

[85] *Arkansas: A Guide to the State*. Compiled by Workers of the Writers' Program of the Works Progress Administration in the State of Arkansas. Original publication in 1941, with a new introduction by Elliott West (Lawrence: University Press of Kansas, 1987), 65.

manufactured products to sell to the outside world. Here in Dyess your principal cash crop will be cotton. You will produce in this community between six and eight thousand bales of cotton a year. That will bring you in around a half million dollars of new money every year. This amount would be sufficient for you to make your land and equipment payments and to buy most of the things you need from the outside." Westbrook eventually got to the subject of most interest to residents — purchase of their homes and land.

"Now as to the terms. First, the corporation will have each tract of land and its improvements appraised. The basis of appraisal will be: Earning capacity for the householder. That is, consideration would be given to the value of the place as a home as well as to the ability of the land and improvements, with ordinary work and care, to produce a living and permit the making of payments. After the appraised value is determined as indicated, the corporation will offer the house to the family living there at that value less 10 percent. We are making this deduction to the family living there not as any special mark of favor, but because we think it would cost about that much to get someone else established there." Westbrook apparently anticipated that some colonists would dispute appraised values, claiming they were too high, so he warned that valuations would not be negotiable unless property failed to sell. "The sales contract between the corporation and the purchaser will provide for the following:

a. A nominal cash payment of $1.00.

b. As now being considered, [a loan] amortization over a 30-year period at Federal Land Bank rates. A $2,500 place would require annual payments of $150. Probably we shall make arrangements so that payments might be completed in 20 years.

c. Provision would be made for monthly, quarterly, or semi-annual payments at the purchaser's option and credit given for interest saved.

d. An emergency clause would provide for the deferring at the discretion of the corporation all or any part of any annual payment and extending the loan accordingly. This clause would be effective in cases where the corporation agreed that failure to make a payment was due to causes not the fault of the purchaser.

e. A restricted sale clause would prevent the purchaser from selling his home to anyone but the corporation until he had fully paid for it. The purpose of this clause [is] to prevent speculation and to give the purchaser's family the security of a home that cannot be bargained away."

Westbrook chose several opportunities throughout his speech to push the virtues of his cooperative concept. "Co-operation must be our watchword. Co-operative principles must be rigidly adhered to, and co-operative procedures must be developed and perfected in such a manner that they will form patterns for the use of the millions of good people who live in our country today without security and without adequate pay for their efforts." He stressed that "in this community we aim to apply the principles of co-operation to all our important activities and at the same time to preserve and develop that spirit of competitive individualism which seems necessary for the highest achievement in any line of human endeavor." Competitive individualism often thwarted Westbrook's efforts to create cooperatives at Dyess, and he admitted that "until recently co-operatives of any kind have had mediocre success." According to him, the end result of colonist efforts would be "the most prosperous community in Arkansas or any other state. There will be no rich people here, but everyone will be well-to-do."

In order to achieve efficiencies Dyess administrators organized the colony into three sections for management purposes with an assistant farm supervisor in each. ARRC's farm management department in 1936 established colony polices. They included a demand that Dyess farmers till their own land for maximum yields before working on community projects. That year colonists planted 4,648 acres: 2,969 in corn; 1,031 in hay; 307 in cotton, and the balance in sorghum and other crops. After harvest, many colonists worked to clear another 3,000 acres and earned $15 for each cleared acre. Despite a lack of substantial support from residents, Westbrook established Dyess Colony Cooperative Association (DCCA) to represent "all enterprises on the colony."[86] DCCA oversight included the blacksmith shop, café, laundry, store, hospital, furniture factory, cannery, service station, shoe shop, cotton gin, garage, sorghum mill, and a printing business. Operating funds and management were to come from colonists.

When legal advisor Milton Colvin completed his draft of documents pertaining to this cooperative, the articles of incorporation contained a deal breaker for many Dyess farmers. Violations of obligations to the cooperative would be "a cause for eviction from the land, [and] it would be possible to [have] this and other conditions continued into the final deed of the property to the colonist when it is paid for."[87] Colvin explained this in a letter to Westbrook dated October 20, 1936. Despite Dudley's efforts to sell his cooperative concept many colonists weren't buying it for a variety of reasons, and their negative response brought what appeared to be retaliation from administrators. A memorandum given to applicants dated December 17, 1936, stated that prospective colonists cannot "buy land unless [they are] willing to subscribe to the cooperative association and conform to the rules and regulations thereof."[88] Growing discontent with Dyess management and policies prompted Westbrook to pay a visit, but he didn't change many minds. As a result, several industries and service enterprises continued to do business on their own and not as part of a cooperative.

---

[86] Hayden, 35.

[87] Articles of Incorporation, Dyess Colony Cooperative Association, October 1936. Everett Henson Collection, Dyess Colony Archives, Arkansas State University.

[88] E. S. Dudley, Memorandum to Dyess colonists, Dec. 17, 1936. Everett Henson Collection, Dyess Colony Archives #2016-33-024, Arkansas State University.

A New Deal in Dyess

Setting aside their differences, Dyess residents held another celebration in October 1936 in conjunction with the second anniversary of arrival of the first 13 pioneers. An October 23 issue of the colony newspaper reported on festivities. "Beginning with regular Sunday services at the community center the program moved through several phases and closed with a historical pageant depicting the progress of Arkansas from the time DeSoto crossed over from the Chickasaw Bluffs to set foot on virgin land to the present when Dyess Colony emerged from the woods." One colonist had something personal to celebrate that day. This *Colony Herald* carried a story headlined, "Deatons Develop Lovely Home From Log-yard and Former Mud-hole." It indicated efforts made by many colonists to improve their homesteads. "When Deaton settled on his place in March of this year there was water nearly a foot deep around the house. Old logs and stumps covered the ground, blackberry briars ran in riotous confusion all over the place. It was a discouraging sight, but Deaton did not mean to be beaten. He began by making drainage ditches and pulling stumps. He plowed and harrowed the yard, hoed it, raked it, filled the mud holes." Deaton's improvements included building a picket fence with discarded mill lumber and a row of lattice along both sides of the house. "Passing the place one is forced to stop and admire its beauty."

Celebrations were a big part of life in Dyess. Above, families gather in colony center for second anniversary festivities. National Archives, Records of the Work Projects Administration. *Below, children participate in a parade.* Courtesy of Jane Moore

12
BIG RIVER
BLUES

The future looked sunny at the end of 1936 for Dyess farmers, but dark clouds moved in carrying too much rain with them. The 1937 flood that struck in January brought misery and destruction to a great many Delta families, including almost everyone at the colony. Mississippi River tributaries contributed mightily to this catastrophe, and much of Northeast Arkansas' damage came from the Tyronza River. The river reached its highest level in recorded history. While this disaster unfolded January temperatures plunged to the low teens. In flooded areas displaced people lacked adequate protection from the cold and ice storms. Food ran short, and ominous cases of pneumonia appeared. One victim described the terror of living near advancing flood waters. "You just can't sleep in a house — even if it is your house — within four blocks of a wild, deep river which threatens to break through weak, soggy levees into your house. You lie awake because you are afraid of the worst that would come with raging flood waters. You imagine you can hear alarms from the guards on the river; you think you hear autos speeding through the streets to the levees to empty [their] men into a crew [to sandbag a levee]."[89]

The flood became the worst natural disaster to befall Dyess during the 1930s. Dudley wrote a lengthy memorandum to Westbrook dated March 19, 1937, outlining the harm it caused. "It rained almost daily during the early part of January, these rains becoming excessive about January 6, 1937, as will be shown from extracts from our Colony

---

[89] Don Hamilton, "Jonesboro and Arkansas During the Flood of 1937," *Craighead County Historical Quarterly*, 9:1 (Winter), 5.

Diary." Dudley recorded the weather each January day. "During the night of Jan. 16 the Tyronza River and Drainage Ditch 40-B overflowed their banks, water covering a small area in the vicinity of the Old Center and across the road just south of the Community Center. This water covered a number of yards, but did not get into any of the houses and remained through the 17th and 18th. It started raining again on the morning of the 20th and continued throughout the day." On January 21 the river again rushed over its banks. "I was called out about that time, made a trip to the north end of the colony and found practically the entire northeast corner of the colony to be inundated. This made it necessary to immediately bring the families out of that section of the colony to higher ground. The work of bringing the families out started before daylight, and by night we had them established in the Community Center." By evening on the 21st "it was necessary to evacuate other parts of the colony, and we had between 700 and 800 people at the Community Center who were forced to leave their homes." By noon on the 22nd approximately 1,500 people had taken shelter at the Community Center.

"Ground immediately around the Community Center [became] about the only place on the colony that was not covered with water." So many people packed into a small space with chilling temperatures, sleet, and freezing rain threatened to cause major health problems so administrators decided to evacuate residents. At first about 500 colonists received money and transportation to homes of relatives and friends living in safe areas throughout the region. However, with no electricity or clean water, with no sewage system or telephone service, with rising water and shrinking supplies, Dudley decided to evacuate the entire colony. "Just as soon as it was decided that evacuation would be necessary, word was sent to the people throughout the colony asking them to scaffold or stack their belongings on tables. This was done in most cases, and there was very little loss of personal property as a result of the water."

Transport through Memphis to the U. S. Army's Camp Pike in Little Rock delivered families to safe ground. There they found mess halls and medical services provided by doctors and nurses transferred from the colony. Most of the ill had colds, a few influenza, and about a dozen suffered from pneumonia. The flood caused no injuries and only two deaths, one from flu, and one from pneumonia. "The water started receding about January 30, and by February 3 the roads were all dry and water running off the farms through drainage ditches. We started moving people back to their homes on February 8. We continued returning approximately 170 people per day from that time on."[90]

Ray Cash and his oldest son Roy never left their home. They remained at Dyess with about 30 men in order to protect property as best they could. Johnny wrote two songs about the experience—"Five Feet High and Rising" and "Big River Blues." In 1980 Ed Salamon interviewed Johnny on his radio show about the origin of these songs and got the country music star's account of the flood. "It had been raining for days and days, and we'd heard the news reports every night. My daddy would turn on that battery radio to get the 8 o'clock news, so we knew that the Mississippi River, which was only six miles from us [actually 17 miles] had passed flood stage, which meant it was up to the levees. Quietly and calmly everybody in the community packed up and got ready to go, except my daddy. I remember him taking a yardstick and leaning over the edge of the front porch as the water had covered our field and was rising up on the steps of the house, and he was measuring how deep that water was. I was only five years old so I don't remember all of it, but I remember him saying one day that we just got word that the bus is going to be here in 15 minutes to take everybody to the train to take us to the hills, and we've got to get ready in 15 minutes."

Ray used a boat to ferry his family to the road where a bus waited. "So we got on the bus—my mother, brothers, and sisters—and I remember [riding] through the water to the railroad station where we got on the train, and I remember part of the trip on the train. There was a man on the cowcatcher of the train with a long pole feeling in front to see for sure that there weren't any logs that had floated over the tracks

---

[90] E. S. Dudley, Letter to Col. Lawrence Westbrook, March 19, 1937. Record Group 96, Farmers Home Administration, AK-80, National Archives. Copy in Dyess Colony Archives, Arkansas State University.

that might wreck the train." During the second night of this journey they pulled into Pine Bluff, Arkansas, and reached high ground. The Cash family travelled farther inland and moved in with Carrie's parents. "We had no news from my daddy for three weeks," Johnny said. "My mother had no idea if he was alive or dead. One morning I woke up at my grandfather's house, and I heard my daddy talking to my grandfather about the water: that it had washed away the beehives; that they lost the cow—she drowned; that he had opened all the doors and windows of the house; let the chickens in; even let the pigs in the house so we could keep them you know because the water had gotten right up to the floor." Though water stopped rising, the Cash family stayed another week on high ground before returning to Dyess. "We went back to the most terrible mess you can imagine. The couch was covered with eggs where the hens had laid. You can imagine what kind of mess [the pigs] made all over the house, but we still had them. The beehives had floated off. The cow was gone, but we had the chickens and the pigs." They also had about six inches of "rich, black river dirt covering the whole place that the flood water brought in." That year they made the best cotton crop the Cash family had ever harvested.[91]

The flood caused some colonists to desert Dyess forever. Janice Wise recalled its effect on her family, including a new baby. "It was raining, sleeting, freezing. Daddy took us to the center. He and his uncle went back to the house because Daddy had a World War I bond there and he didn't want to leave it. We worried and worried. Finally, they got back. When we loaded on that box car late in the evening . . . they wanted Momma to go on the hospital car that had a heater in it. She wouldn't leave us [five] children. So, she had a few clothes in a pillow sack and she sat on that pillow sack on the floor" and held her six-day-old baby. "My older sister held a little two-year-old brother." Wise said that her grandfather lived in Columbia County and met the family in McNeil, Arkansas. "We lived with them two months, I guess, before we got another place to live. He gave daddy some land that the sawmill had clear cut, and built a barn first, and put some partitions in

---

[91] Johnny Cash, excerpt from interview conducted by Ed Salamon, program director, WHN Radio, New York. *The Johnny Cash Silver Anniversary Radio Special*, aired on the Mutual Broadcasting System, July 4, 1980.

it. We lived in that barn until the next year. It had no insulation. I don't know how we kept from freezing." After the floodwater subsided her parents returned to Dyess to pack, and they moved back to Columbia County.[92]

---

[92] Janice Owen Wise, Oral History Interview with Emmett Powers, September 18, 2013. Dyess Colony Archives #2018-10-018, Arkansas State University.

During the 1937 flood the Dyess Center, above, was the only high ground in the colony. Blytheville Courier. *Below, travel through Mississippi County became treacherous during the flooding.* Photo by Curtis Duncan

13
BETWEEN TWO FACTIONS

By mid February 1937 most colonists had returned to their homes and begun cleaning up the mess and correcting structural damage to their houses and businesses. Dudley summarized his views about the disaster and aftermath in a message to Westbrook. He stated that the majority of colonists were not discouraged and were deeply appreciative of the way they had been cared for. "We have not lost more than twelve families who in my opinion are leaving because of the overflow."[93] Two department of agriculture economists who studied the subject found that "between May 1936 and April 1938, 194 families moved away from Dyess. The most frequent reason given by them for leaving was general dissatisfaction, 37.7 percent, out of which 5.5 percent specifically directed their dissatisfaction against high water and mud."[94] Though the colony had substantial flood damage, administrators proposed to bear the expense of restoration without outside assistance. But their plan had to be abandoned as costs mounted. In October the Red Cross granted colony managers $19,265 to repair damage caused by the disaster, less than one half of the projected cost. Still, Dyess farmers planted about 3,000 acres of cotton that year, which by the end of the harvest yielded more than 1,500 bales.

Despite difficulties brought on by man and nature, that following summer about 30 families received homestead deeds for purchase of their farms. Repayment terms included a 30-year

---

[93] E. S. Dudley-Col. Lawrence Westbrook, March 19, 1937.
[94] Hayden, 52.

amortization and annual payments averaging about $120. Jack Bryan reported in a November 26, 1937, *Memphis Press Scimitar* that colonists had much to be thankful for. "Instead of fleeing government tyranny like the early [Colonial American] colonists, these people have been brought to a new land by a benevolent one and supplied with leadership to get them out of the Depression." Another florid write-up appeared in the *Arkansas Gazette*. Frank Newell referred to Dyess as a "new American frontier in the fullest meaning of the phrase. Men whose speech is studded with pure 'Anglo-Saxonisms' are foraging a civilization out of a jungle and wilderness."

Unfortunately for colony leaders lauded by these reporters, state political battles in 1937 damaged Dyess. Historian Don Holley explains the problem. "Arkansas politics [during the period] was a contest between two factions, the statehouse crowd led by Governor Carl E. Bailey versus the federal crowd," which included U. S. congressmen and federal office holders. Sharp, who replaced Dyess after the founder's death, "was caught in the middle."[95] At that stage the community no longer possessed the political cover that Dyess had established with Futrell, and animosity increased in 1937 when Bailey lost a U. S. Senate race to John E. Miller, a federal faction favorite. Though Sharp ostensibly remained neutral, Bailey forces claimed that his WPA political operatives helped defeat their candidate. Among several acts of revenge, Bailey complained that the colony failed to pay $33.00 of franchise taxes. Futrell, now the colony's attorney, argued that being a federal entity, DCI did not have to pay the tax. He also pointed out that the state's tax commissioners previously accepted this argument. Aware that he had become a target and not the colony, Sharp asked Hopkins to relieve him of responsibilities there in order to protect the community from further political damage. Prior to his resignation Sharp obtained a new Arkansas charter that established Dyess Rural Rehabilitation Corporation (DRRC) as successor to DCI. The new organization's goal was "To rehabilitate individuals and families as self-sustaining human beings."[96]

---

[95] Holley, 210.
[96] Ibid., 214.

For purposes of colony oversight its administrators established leaders, associations, and boards to help direct community affairs. An executive administrator monitored official leaders. These persons included education and recreation leaders to oversee the school, churches, library, and social clubs. The education leader served on the school board with a superintendent and members from four wards. This board had authority to hire teachers, approve building programs, control finances, and supervise other activities. A planning board consisted of an executive administrator and an economic leader plus five other persons. This group bore responsibilities for directing different enterprises such as crops to be grown. A marketing association handled commodity sales at the end of a production cycle. This group also oversaw purchase of seed and quality control of crops produced. A home improvement association sought to raise colonists' standard of living. Its members pushed for yard improvements where needed, family meals that met nutritional guidelines, and appropriate furnishing of homes. Such monitoring programs for virtually every personal and professional matter incensed some colonists, but animosity may have been driven somewhat by the nature of Dyess settlers.

In his appraisal of resettlement colonies, C. H. Hammar Jr. pointed out that the typical personality of colonists made transition to a new way of life difficult. He concluded that those "needing resettlement include the most adventuresome and individualistic people in the population." They most likely would resent the administration's interference in their lives, which made friction inevitable. Hammar observed that Dyess "can by no means hold all its clients or settlers." He suggested that this colony would "have a higher turnover rate" than others.[97] A Dyess case worker interviewed at least one disgruntled colonist with an independent streak and included his complaints in a report prepared for the case worker's supervisor. The interviewer called this man Mr. G., who "stated that when he left [his home] county he had been a satisfactory rural client for one year, that he had repaid his loan, had a credit of $43.50 to his account, [and

---

[97] C. H. Hammar, "An Appraisal of Resettlement: Discussion," *Journal of Farm Economics*, 19:1 (February 1937), 203.

owned] a pair of oxen, a cow, chickens, bees, and 1,000 cans of vegetables and fruits." Mr. G. claimed that he needed 40 acres of ground rather than 20 acres offered by Dyess administrators. "His son and wife were listed as members of his household, and they could easily work 40 acres. He said that he asked to [bring] his oxen and cow, but was told that mules would be used to work Dyess Colony [land]. Also, he was told that at Dyess Colony they would be furnished a milk cow; therefore he left his cow with his daughter. After his arrival at Dyess he was assigned to a small house and only 20 acres."

The report concerning Mr. G. also revealed uneasiness about colony homestead costs. "The purchase price of the home and land had been explained to him as probably not exceeding $100.00 per acre and not less than $80.00. He was not satisfied with this, saying this should have been stated in the contract: that in 12 months the people in charge could be transferred, and that he would have only the verbal statement of the person in charge, and that he did not wish to assume a responsibility of which there was no definite understanding. Mr. G. stated that his neighbors there were also dubious of the contract and agreement, but they were afraid not to sign, fearing they would be cut off, and that they had no other place to go." Mr. G. claimed that his neighbors "were also dissatisfied due to the fact the rules and regulations required of clients were not set forth, approximate amount of purchase price not stated, amount of work stock, etc., not included, and arrangements for marketing surplus production not set forth. Mr. G. said he wished to have a definite understanding of just what was expected of him. He said he did not like the idea of waiting two years to have the final papers drawn up. Mr. G. stated that [he] had been at the colony three months before he saw a copy of the contract [and] that all this time he was uneasy and dissatisfied." According to the report, "Mr. G. stated that when he left the colony he asked for a receipt for the things the colony repossessed; this was not furnished him. He said he was told that his work would be appraised, his account cleared, and a copy sent to the state department of county relief and a copy to him. This he has never received."[98]

In addition to colonist dissatisfaction about land prices and uncertainty about terms of sale, they opposed use of chattel mortgages to secure debts owed to the colony store. This meant that debtors who failed to pay bills when due might lose their assets through foreclosure and repossession. Farmers also complained about excessive prices for crop inputs at the store. For example, at Dyess they paid $21 for a ton of hay when purchased with doodlum. The same item cost only $12 when bought with cash. "Many other articles [were] similar in ratio."[99] Displeasure with prices may have stemmed from the fact that many Delta planters enforced a similar arrangement with tenant farmers, making them purchase products from a company store at inflated prices if using doodlum. To be victimized at Dyess in this manner would have been especially irksome for colonists.

---

[98] Dyess caseworker report. Everett Henson Collection, Selection Process, Dyess Colony Archives, Arkansas State University.
[99] Hayden, 35.

The cotton gin, above, and community cannery, below, were among Dyess cooperative enterprises. Above photo, Everett Henson Collection, Dyess Colony Archives, Arkansas State University; below, Library of Congress

14
A Dog
Named Jake

Jake Terry and E. S. Dudley enforced many unpopular measures at Dyess, and they threatened farmers who wouldn't sign chattel mortgages with penalties such as denial of WPA work. Ongoing discontent with colony management practices dated back to the community's first days. In an RA report prepared by Mary E. Johnson in 1936 she claimed that local management, including Dudley and Terry, kept their charges in line by "control of furnish, land, and houses as well as equipment," a common arrangement in the plantation system.[100] Terry became such a divisive figure that Sharp complained about him to Westbrook in an April 8, 1938, letter. "In checking into the affairs at the colony we find, invariably, that the colony families resent him, and Mr. Dudley has just revealed to us the fact that a part of the administrative force, including Mr. Runyan, are at outs with him. Just recently a group of women went to Mr. Dudley in regard to their farm problems, and when he tried to get them to go to Mr. Terry they frankly stated that they and their husbands knew that if they talked to him they would get cursed out." Apparently Terry refused to be managed by Dudley, his supervisor, claiming that he answered directly to the board of directors and not Dudley, an incorrect claim. Colonists and administrators who worked with Terry agreed that he had a "bad

---

[100] Mary E. Johnson, "Preliminary Report on Community Development Policy," July 3, 1936, National Archives, Record Group 96, Farmers Home Administration, AK-80. Copy in Dyess Colony Archives, Arkansas State University.

attitude."[101] Ray Cash reportedly disliked Terry so much that he named the family dog Jake.

Some grievances proved to be beyond the power of administrators to cure. Poor prices in 1938 led Dyess farmers to seek an increase in their cotton acreage allotments. They claimed that existing limitations reduced acreage to an unprofitable level. This meant that for many colony farmers growing cotton was not worthwhile in 1938 because they were not allowed to plant sufficient acreage.[102] Dudley's attempts to switch farmers to vegetable crops met resistance for several reasons, one being a cooperative approach to marketing. Another dispute concerned dismissal of several popular commissary employees. According to management, they failed to charge for everything that went out the door. Turmoil surrounding these and other disputes resulted in 100 disgruntled colonists signing an affidavit of protest. Though his motivations remain unclear, an outsider took it upon himself to lead a revolt. Paul Finch farmed land adjoining Dyess, and after meeting with angry colonists he complained to Arkansas officials about colony administrators. He also sent two letters to the *Memphis Press-Scimitar*. It did not help Finch's case that in one letter published April 20, 1938, he commented on issues as diverse as the Confederate bombardment of Fort Sumter and U. S. postal policies in the mid-1800s.[103]

In a second letter published on April 21 Finch claimed that "there are anywhere from 100 to 150 houses vacant in Dyess Colony, and 98 percent of the balance are dissatisfied and would leave if they had a place to go where they could make a living." His call for new management reflected the views of many colonists who claimed that administrators treated them like children. When Finch held a meeting attended by 45 Dyess residents the group proposed three changes: establishment of a 12-man committee to hear disputes between

---

[101] Floyd Sharp to Col. Lawrence Westbrook, April 29, 1938, Work Projects Administration, Special Collections, University of Arkansas Libraries, Box 10.

[102] Hayden, 65.

[103] Work Projects Administration, Special Collections, University of Arkansas Libraries, Box 10, File 87.

colonists and administrators; replacement of the existing DCI board of directors with a new board comprised of three colonists, and an increase in furnish. After being contacted by Senator Hattie Caraway about Finch's charge of improper federal expenditures at Dyess, Sharp wrote her on June 6, 1938, with blunt advice. "I strongly urge that you not answer his letter. If you write him anything he will give a speech and interpret it to suit himself. Last Friday night he was on the colony making a speech, and it ended up in a free-for-all fight."[104]

Finch's publicized complaints caused some colony advocates to refute his criticism. Soon after his letters ran in the Memphis newspaper Velma Bullard, whose parents lived in Dyess, wrote a letter to the *Memphis Press-Scimitar* published on May 2. She blamed outsiders and lazy colonists for problems at the colony. "I know plenty of people you couldn't get to move out. Dad says it's a real chance for anyone that wants a home and is willing to work for it." Discontent prompted administrators to make some changes. They implemented more training for workers involved in retail enterprises. DCI arranged to increase cotton acres and enticed 150 colonists to join a cooperative to gin, store, and market cotton for its members. In addition to public relations problems colony officials had other deficiencies to worry about. At the start of 1939 about 150 farm families failed to meet their financial commitments.

DCI chose to address this and other problems in a memo from its board of directors to colony families. It notified them of several changes to come:

(1) Boards of advisors would manage each unit controlled by DCI, such as the hospital, gin, etc. They would "work under the direction of" DCCA, which encouraged individual boards to meet with unit managers on a monthly basis.

---

[104] Floyd Sharp to Senator Hattie Caraway, June 6, 1938. Work Projects Administration, Special Collections, University of Arkansas Libraries, Box 10, File 87.

(2) At a joint meeting of Farmers and Exchange and Loan Company board members and DCI's board they decided that crop loans for 1939 would be financed by the loan company using different credit standards. "In order that each family may have information now as to its available financing for 1939, please be advised that loans on an acreage basis will be abandoned." Starting in 1939, loans would be made on the basis of a borrower's ability to repay it from cash crops. This important change tied financing to production, a practice consistent with commercial credit sources. DCCA members could borrow funds at five percent interest. Others would be charged ten percent, perhaps punishment for those who opposed cooperative membership. "Crop loans must be secured by a first mortgage on the crop as well as a first mortgage on all work stock, cattle, and heavy farm tools."

(3) Perhaps the biggest benefit for cooperative members came in the form of debt forgiveness, spelled out in the memorandum. "To all families who are [qualifying] members of the Dyess Colony Cooperative Association . . . Dyess Colony Incorporated will adjust individual accounts on the following basis: Item 1. All subsistence debt for the years 1934, 1935, and 1936 will be entirely written off. 2. One half of the subsistence debt for 1937 [the flood year] will be written off." These write-offs created an enormous benefit for eligible farmers.[105]

This memo also addressed other major changes. After extolling many achievements of its hospital, administrators announced that "Dyess Colony Incorporated cannot continue to subsidize its operation as it represents too heavy [an] expenditure against our remaining capital funds. You are therefore advised that effective July 1, 1939, Dyess Colony Incorporated will discontinue the operation of the

---

[105] Floyd Sharp, "Memorandum to All Colony Families," Dec. 30, 1938. Everett Henson Collection, Dyess Colony Archives, Arkansas State University.

hospital. We are hopeful that your cooperative will be able to work out some plan for its continued operation whereby it can be made self-sustaining. Dyess Colony Incorporated will make the facilities of the hospital available to your cooperative association without cost and will work with you in every possible way to assist in the solution of this problem." With respect to the school, "Dyess Colony will continue to subsidize its operation at the present level through the present school term. You are advised, however, that for the school year 1939-1940 it will be necessary for the expense of operation to be materially reduced, and any subsidy expended on this unit must come from profit-making units.

"Let us again remind you that Dyess Colony Incorporated is anxious and willing to turn these properties completely over to the cooperative association as soon as competent management can be developed and the enterprise worked up to a point where substantial payments have been made, or will be made, to retire the cost of the units."[106] Perhaps administrators anticipated negative reactions to their policy changes because this memo announced a 60-day move-off policy. "It is realized that we have a few families who, on account of ill health or for various other personal reasons, are not satisfied at Dyess Colony. To any of these families the board of directors of Dyess Colony Incorporated is announcing now that it will assume the responsibility and cost of moving them back to their home county or to a location an equal distance from Dyess Colony, at any time within the 60-day period beginning January 1, 1939, and ending March 2, 1939. We feel confident that our mutual problems will be successfully solved and we are looking forward to the expansion of the present Dyess Colony Cooperative Association."[107]

Despite some peace offerings, skirmishing between farmers and administrators continued. On March 29, 1939, an *Arkansas Gazette* reported a meeting attended by 275 colonists at Osceola's courthouse to protest "exorbitant charges for land they had bought." At the end of the meeting they sent a telegram to President Roosevelt and Arkansas

---

[106] Ibid.
[107] Ibid.

Governor Bailey calling for assistance. According to this newspaper account their telegram disputed the cost of land. "The records of this county show that this land was purchased at $2.50 per acre. Our contracts call for its sale to us at the actual cost plus improvements, but officials now insist that we agree to pay from $75 to $100 per acre, many times the value of the land. Unless you can prevent this repudiation we will lose four years labor and our last chance for a home. Please help us!" Pricing land had always been a problem caused by imprecise terms. The fact that commitments sometimes came in the form of oral representations rather than in writing also bothered colonists. Regardless, management faults appear more the result of administrative bungling than attempts to cheat farmers.

Another press account announced difficulties of two protesters. An April 1, 1939, *Arkansas Gazette* explained that Dyess administrators were in the process of expelling A. J. McCravin and S. B. Funk from the colony. "A Blytheville lawyer has been hired by the colony to assist J. M. Futrell, former governor and now attorney for the colony, in trying unlawful detainer proceedings which were started more than a year ago." In a letter from Sharp to the WPA's Washington office dated April 24, 1939, he wrote that these two farmers refused to accept deeds to their homesteads "at the regular established price."[108] In Mississippi County Chancery Court both colonists "protested the cost of their land, alleged abuses, and other matters." As a result, colony administrators began an eviction action. According to the *Arkansas Gazette*, colony representatives filed papers on March 17, 1938, demanding that "Mr. McCravin and Mr. Funk give up goods and deliver property within three days because of having been incompetent and insufficient in work." Seven days later a judge ordered them to "appear in court immediately." They did and made bonds of $400 each. The two remained on their farms, but the colony store refused to provide credit. They obtained credit elsewhere and continued to farm "without knowing their standing one way or the other," according to the *Gazette*. The Arkansas Supreme Court settled this matter. It upheld authority of

---

[108] Floyd Sharp to WPA Office, Washington, DC, April 24, 1939. Work Projects Administration, Special Collections, University of Arkansas Libraries, Box 10, File 88.

managers to eject colonists after an unsatisfactory probation period and sustained the ouster of Funk and McCravin.

Ongoing controversy about valuations led colony administrators to retain Russell Brown & Company, Certified Public Accountants, in Little Rock to analyze costs associated with individual home sites. Their 1939 report revisited audited financial statements from February 29, 1936, to provide a detailed breakdown of steps taken to establish property costs. "In compliance with your recent request we are submitting herewith a statement of our procedure in determining the cost of each of the various farm houses, barns, poultry houses, and house sites at the Dyess Colony, particularly with reference to House #8 occupied by S. B. Funk." It is important to note that this audit presented valuations determined by independent accountants. So one may conclude that the firm's numbers are probably accurate and relevant to disputed valuations.

Their report explained that "During the construction period a separate cost sheet was maintained for each house, barn, and poultry house constructed." It provided a detailed breakdown of costs used to establish purchase prices for colony homesteads. "While we were making our audit of Dyess Colony, Inc. as of February 29, 1936, we [determined] the average cost of each house, barn, and poultry house and established this average as the basis for future sale rather than the individual cost of each of the various houses." According to this report the average cost of a "five-room Type E house such as house #8 was $1,963.50. The average cost of a Type E barn such as was constructed on this place was $265.56. The average cost of a poultry house was $62.82. This will make the total cost of the improvements placed upon the land $2,291.88."

The contract included legal expenses, cost of quit claim deeds, delinquent tax payments, and surveying fees, all of which totaled $382.88 for the 38.99 acres in Tract #8. The cost of drainage ditches directly benefiting the property came to $279.95. "In view of the fact that the various tracts were to be sold as cleared land, $15.00 per acre was included in the cost price and set up in a reserve for clearing. This added $584.85 to the cost of the land." According to these accountants, "the foregoing amount [established] the total cost of the land in Tract #8 [to be] $1,247.68, which added to the improvements will [bring] the total cost of Tract #8 [to] $3,539.56." Funk received a $273.90 credit, roughly $15.00 per acre, for land cleared by him. "This credit reduced the above cost to $3,265.66, whereas the sale price to Mr. Funk was $2,999.52." Therefore the sale price seemed reasonable despite Funk's protests.[109]

---

[109] Audited Financial Statements, February 29, 1936, and 1939 report, Russell Brown & Company, Certified Public Accountants. Everett Henson Collection, Dyess Colony Archives, Arkansas State University.

The Dyess school was the source for most entertainment in the colony, including a full athletic program for males and females. Above, Everett Henson Collection, Dyess Colony Archives, Arkansas State University; below, Courtesy of Jane Moore

15
Good Days
and Bad Days

In the fall, when evening air was a bit cooler, farmers emptied the last full cotton-picking sack into a trailer. This ended another harvest. A tractor hitched to the trailer pulled it to a gin and across weight scales. The cotton would be sucked through a wide metal tube into gin stands that separated seed from lint. Ginners compressed lint into 500-pound bales and sold them to cotton merchants. Gin owners then cut checks to growers for their share of the proceeds. From this money farmers paid off debts that accumulated during the year. If they had money left over, a sense of satisfaction, and sometimes euphoria, possessed the family.

Everett Henson recalled a family ritual after getting paid for their cotton. "We'd come to town, to the co-op store. Our favorite meal was bologna and cheese and all the crackers you [could] get out of the barrel. They'd put butcher paper out there and lay that stuff on it. And we'd stand there at the counter and eat and that was our meal. Sometimes we had to go outside and get us a drink out of the faucet, and sometimes we'd have enough money for a soda pop."[110] His brother A. J. added that their father would usually buy his kids ice cream cones when they came to town. "I thought dad didn't like ice cream, but I found out years later that he didn't have enough [money] to buy himself one, too, so he bought the kids one and that was it. Times were really tough.[111]

---

[110] Henson, Everett.
[111] Henson, A. J.

Everett Henson also remembered his father building a swing in their yard by "connecting four trace chains together, tying them in a tall tree, putting some straw in a tow sack, and then fastening it to the chain. We would climb a ladder leaned against another tree and swing off. Someone would climb the ladder and jump on the swing, and both would swing together. All the kids on Road No. 10A and 10 would come and swing on Sunday afternoons."

On Saturday nights "people gathered in the circle [at the colony center]. Horses and mules were all tied down. There were two or three rows of hitching posts for people to tie off their mules and wagons. The administration asked people not to tie mules to trees because the animals ate the bark. On the right side of the administration building was a building that had a huge café inside with 90 seats. This building later became a bank, then it had the picture show." The Hensons made frequent trips to the colony store. "The commissary store had just about everything in it. No merchandise was put out except the sacks of potatoes and the cans of kerosene. If you needed something a clerk handed you the supplies. Everyone was encouraged to shop there. The store belonged to the co-op, so patrons received a doodlum book and were able to trade doodlums at the store." His life differed dramatically from what people endured outside the colony. "We lived next to the Lee Wilson Plantation, and the people between Dyess and Marie [a small town nearby] were on the rehab [welfare] and lived in houses that weren't as good as the Dyess houses. I played with these kids and went to the picture show at Marie on Saturday nights, and we always had a good time together. I don't know how they felt about the people at Dyess Colony. I would think they would have been jealous of the pretty white houses we lived in. Their houses were rough sawed and weren't sealed on the inside. They were renters, sharecroppers, or day laborers."[112]

Many of Malcolm Gordey's memories included pranks and entertainment. "If there was any free time we played cards, checkers,

---

[112] Everett Henson, "Memories of the Dyess Colony," *The Delta Historical Review*, 2:1 (Summer 1990), 6-10. Dyess Colony Archives #2018-14-004, Arkansas State University.

or dominoes." Members of the Gordey family read anything available, including magazines such as the *Saturday Evening Post* and *Colliers*. Gordey often checked out books from the library, and two of his favorites were *Gone With The Wind* and *All Quiet On The Western Front*. He remembered listening to the radio when Orson Welles presented *War Of The Worlds* with such realism that it terrified listeners throughout the nation. They believed that alien creatures had actually landed and attacked America. "Next day many of the people said they did not know until the program was over that it was not a real happening."[113] Gordey's reflections about school days tended to be humorous. At that time vehicles transported older students first to Keiser, then Wilson or Shawnee. "We rode in a truck with a tarp over it. In rainy weather it was fine until the truck stopped; the fine mist of rain and mud flew all over us. We needed a bath before classes, but had to make do at the wash basin."

The school eventually transported students in the "cracker box," a vehicle similar to a bus. "It was fine in the winter, but as spring came it began to get hot. We tried to open the front and back doors, which helped, but soon we felt we needed more air. Suddenly the glass windows began to disappear, and even the seats in the middle of the bus got tossed out." The cracker box became the scene for many pranks, including use of slingshots to hit passing people, cars, and cows. One man whose truck was struck by a missile took after the bus. Fortunately for the students, "when we went over the bridge to Dyess the truck turned off, and we never heard anything more of it." In another incident aboard the cracker box a girl's slip she made for a home economics project fell out of a box, and a male student "picked it up and waved it out of the back door like a flag. Of course, somebody saw it, told Mr. McGuire, the superintendent, and a school board member or two. I think we were suspected of having an orgy on the trip. I know we were all grilled pretty hard, but they got the same story from all of

---

[113] Malcomb Gordey, Letter to Everett Henson, July 6, 2000. Everett Henson Collection, Dyess Colony Archives #2016-33-181, Arkansas State University.

us. So they quieted down. Some of them acted like we should not even know that women wore slips."[114]

Winford Henry had an unpleasant problem with schoolmates. "I had two boys, they was against me. I carried my lunch to school in a little old bucket, and they would meet me out between the school and the bus . . . and they would jump me and mash my bucket and everything and Burl Smith was one of them. And I went in one night and daddy told me, 'I tell you what, if you come in here with that bucket smashed again . . . I'm going to wear you out.'" The next time the bullies confronted him "I was walking out there with my little Rex jelly bucket, and Burl walked up to me to jerk my clothes and I hit him and knocked him down and I got on top of him, and I was beating the fire out of him, and they finally pulled me off of him, and Burl Smith became one of my best friends I ever had." This friendship apparently flourished despite a terrible accident. "I was going down in front their house in an old Dodge pick-up one Sunday morning and their little sister run across the road to the mailbox. She wasn't but seven or eight year old. I hit her with that truck and knocked her from the right side over to the left side of the road in a ditch. I stopped and run over there, picked her up and . . . laid her in the truck seat, didn't wait for them to come out, and headed for the hospital here at center. And she finally made it, but she was crippled from then on."[115]

During a school trip several students in a truck were injured in a wreck. James Phillips said that "We were on our way to Hot Springs to the FFA [Future Farmers of America] camp and got run over by a big semi-truck. And one boy was killed, and the ones that weren't hurt too bad ended up" with Dr. Hollingsworth. He "painted us up with iodine and all that. And he said, 'Would you like a blow torch to cool that off?'"[116] Another vehicle accident claimed the life of a young boy.

---

[114] Ibid.

[115] Winford Henry, Oral History Interview, Memories of a Lifetime Project Team, October 20, 2017. Dyess Colony Archives #2018-12-005, Arkansas State University.

[116] James Phillips, Oral History Interview with Ruth Hawkins, July 7, 2018. Dyess Colony Archives #2018-33-004, Arkansas State University.

Mary Mauldin remembered that sad event. "There was a rolling store. It was a truck with a big wooden bed, and it would come around about once a week and they had all kinds of kitchen stuff. They had flour, meal and all that kind of stuff. They even had some material for the ladies, and thread. And they had a big chicken coop on the back." People could sell their chickens if they wanted to. "And he always had candy.... There was a little boy that would always run out and meet that rolling store and the man always gave him a stick of candy. And for some reason, I don't know what happened. The man didn't look real good and the little boy got run over by the rolling store."[117]

Though sometimes bad days came and went, Jean Ann Jennings described her life in the community filled with mostly good days. "We never locked our doors and it's hard to explain to the modern folks what it means to have a thin screen door between you and the outer doors. Thin boards and a tacked on screen was all that separated us from whatever was outside. We never thought of intruders. The outside light we had was the moon and the lightening bugs. All families that came here were around the same age, everyone knew everyone, and you even knew everyone's dog's names. We were poor but never worried about our equally poor neighbors trying to steal the little that we had. Everyone was accepted regardless of their shortcomings, mental status or handicaps." Colony children "were free to roam all over, no one had to watch us at all. We could go and be gone for hours and our parents knew that we were safe, that no one would harm us, only help us." Jennings said that in those days "life moved a little bit slower, folks seemed to be a little bit friendlier and people took the time to get to know their neighbors."[118]

---

[117] Mauldin.
[118] Jennings.

Above, Cash colony home in 1936 (either Jack or J. R. standing at steps). **Courtesy of Glenetta Rivers Burks.** *Below, the Cash family ca 1949. From back left, Roy, Carrie, Louise, Ray, Reba, and J. R. (Johnny); front Tommy and Joanne.* **Courtesy of Joanne Cash**

16
A Very
Sad Occasion

Dyess Colony reached its peak in the late 1930s and steadily declined thereafter. A regional director of Farm Service Agency (FSA) agreed to participate in the project if management made some changes. One called for closing out about 200 farms in order to allocate that land to the remaining 300 operations with a minimum of 40 acres each. Small farms had become financially unsustainable. The Bankhead-Jones Farm Tenancy Act authorized FSA to take charge of resettlement operations, but following a troubling status report the colony became Dyess Farms Incorporated (DFI) in June 1940. According to author David Hayden, the corporation had ambitious goals. "Articles of incorporation gave the new company extensive powers at Dyess," including authority to carry on businesses in agriculture, merchandising, and manufacturing. DFI purchased from DRRC about 13,700 acres in the colony for $650,000. The deal excluded the colony's center, amounting to about 300 acres, which DRRC retained for possible development of new businesses.[119] In another attempt to plant seeds of cooperatives in the colony, administrators set up Dyess Cooperative Gin Association, Dyess Cooperative Store Association, and Dyess Medical and Health Association. FSA loaned start-up money to the enterprises but insisted that they pay their own operating costs. Like many cooperatives before them they eventually failed.

Despite a steep reduction of the colony's population, the state's increased from about 1,750,000 persons in 1930 to 1,950,000 in 1940. As

---

[119] Hayden, 96.

private sector employment grew during an industrial ramp-up to World War II the size of relief rolls decreased. The federal government grew less and less interested in resettlement projects and other Depression-era programs as its focus shifted to threats posed by Germany and Japan. The Japanese attack at Pearl Harbor may have been a surprise, but America's likely involvement in a conflict became increasingly obvious during 1940 and 1941. The *Blytheville Courier* often reported preparations underway in area communities such as Dyess, including calls for men to enlist in the CCC. Mississippi County regularly led all the state's counties in enrollment of participants in that agency, and in one year it sent more than 500 men between the ages of 17 and 24 to CCC camps in various parts of the United States. There they went through a training program and received $30 per month and all upkeep. Though a civilian work force, CCC cooperated with service branches to use that federal program as a source for military recruits. Members received information about a National Defense Program and encouragement to join that effort.

When the U. S. formally entered World War II on December 8, 1941, federal control of Dyess operations increased and received a patriotic reception. A national Agricultural Defense Relations agency set up local boards to help businesses meet expanded economic needs. In rural areas these boards focused on increased production of food and commodities required for military support. The colony's manager became Dyess' defense board chairman. A call for support brought community participation in initiatives such as Food For Victory. Organizations as diverse as the PTA and Campfire Girls committed their members to buying stamps and bonds. Elementary school students put up a bulletin board in their school hall with pictures of presidents who led the nation during wars. Colonists increased their gardens and livestock production for the war effort. In a May 22, 1942, *Osceola Times,* Dyess received a compliment for its support of American military needs. According to that edition the community set "a patriotic example that our entire county, or better, our entire state, may well follow." Women made patriotic examples as well as men when they joined military services. In March 1943 Georgia Beatrice Eudy began Women's Army Auxiliary Corps (WAAC) training in Iowa. During

May, Anna Flaherty enlisted in WAAC and commenced basic training. Soon this support began to cost the colony in both treasure and blood. On May 6, 1942, Clyde Williams learned that his son, Willie Lee, died in action at Pearl Harbor while serving aboard a destroyer.

While the colony lost men in battle, the state lost residents. Almost 11 percent of its population looked elsewhere for opportunity from April 1940 to November 1943. Lacking significant facilities for war-related production, the state's two primary contributions became men and food. Despite the need for agricultural commodities, during 1943 almost 20,000 Arkansas families abandoned their farms. They joined millions of southerners who became northerners in search of better jobs and higher pay. Their departure contributed to "one of the most significant demographic shifts in American history."[120] Back on the farm, some New Deal programs and agencies still existed to support agriculture, such as Commodity Credit Corporation loans that put a floor under commodity prices. Federal assistance continued to help sustain a shrinking farm community. Dyess officials had unused houses and barns torn down in order to turn small operations into large operations. Mechanization of production vastly increased the number of acres a man could manage. However, some Arkansas farmers had trouble understanding the uses of new technology. One wrote his congressman about the "new atomic bomb" dropped on Japan. Could the congressman get him one, the farmer asked, "to use in his fields to get out the stumps."[121] Fortunately, practical ideas preoccupied most Dyess folks. Workers tripled the size of barns, repaired houses, installed larger water pumps, and lowered wells to about 50 feet. Though such changes improved life there, the community lost most of its people, businesses, and, some might say, its soul.

Unfortunately, patriotism, hard work, and sacrifices did not protect the colony from loss of federal support. In 1943 the U. S. Congress passed a bill withdrawing resources from resettlement projects. Money and attention were needed elsewhere so FSA began

---

[120] James T. Patterson. *Grand Expectations. The United States, 1945-1974* (New York: Oxford University Press, 1996), 19.
[121] *Arkansas Odyssey*, 475.

liquidating Dyess assets. During this process FSA required that residents agree to new land purchase contracts in order to be eligible for additional loans. Many Dyess farmers considered the new terms unfair. They protested to their congressional representatives, and FSA push back led to eviction notices against 25 farmers by February 1944. Most of them lost their land. In June 1944, 117 colonists attending a DFI meeting approved a liquidation plan that featured several advantages. It voided a previous agreement for DRRC to sell 13,711 acres for $650,000 to DFI. The new plan cancelled DFI debt and deeded land to that corporation. As a condition of this conveyance DFI agreed to sell farms to eligible buyers with FSA financing. DRRC retained control of properties not deeded to DFI. The community store and medical facility both purchased their assets from DRRC and tried to make a go of it, but in a few years they went under along with the hopes of a few Dyess survivors who struggled to stay afloat.

At 7 p. m. on August 14, 1945, President Harry Truman announced the end of World War II. This conflict claimed the lives of many Dyess soldiers, but one notable death during the war occurred in the colony. This tragedy struck the Cash family. In a June 11, 1995, edition of *Parade* magazine Johnny described the death of his older brother and the aftermath. Jack worked Saturdays at the school cutting fence posts for $3.00 a day. On May 12, 1944, Johnny urged his brother to skip work and go fishing with him. "'No, I gotta go. We gotta have that $3.00,'" Jack said. Johnny walked with him a mile to the turnoff to his fishing spot. "Later, my dad came along the road in a car with the preacher. He told me Jack had been hurt. He'd been pushing a post into the table saw, and it jerked him into it. He lived eight more days." Ann Blue remembered Jack and his many contributions to the community. He was a Boy Scout patrol leader until the accident and leader of a Baptist youth group. "I can remember the funeral procession going in front of our farm. It was very sad occasion for the community."[122]

Jack was buried on a Sunday. The following Monday family members walked into a cotton field and began to hoe. But Johnny's mother could not suppress her anguish. "I watched as my mother fell

---

[122] Blue.

to her knees and let her head drop onto her chest. My poor daddy came up to her and took her arm, but she brushed him away. 'I'll get up when God pushes me up,' she said. And soon she was on her feet, working with her hoe." When Jack was a young boy the December 18, 1936, *Colony Herald* published his Christmas wish list. "I am a little boy in the second grade. I want you to bring me a tricycle, a train, and some fruit, nuts and candy. Please do not forget me. Your little boy, Jack Cash." His last wish was granted.

Dyess Colony held great promise for desperate farmers relocated from all over Arkansas for a new start in life. Pictured here, Dorothy Virginia and James Roscoe Phillips, children of Roscoe and Pearl Phillips, in their cotton field at Dyess. **Courtesy of James Roscoe Phillips**

Epilogue

A Poor Man's Best Proposition

When the 20th Century ended less than 500 people lived in Dyess. But many former residents began returning for reunions starting in the early 1980s. They found little but each other and rows of cotton and soybeans. The majority of the project's original structures no longer existed. A few brick ranch houses dotted the area as well as collapsed and abandoned structures. The brick administration building still stood, but little else of the colony center. Dyess had become a ghost town haunting those trying to recapture its past. Though the white brick administration building had been added to the National Register of Historic Places, it looked forlorn and neglected. Everett Henson described it sadly and mentioned that during one visit to the site he found empty beer cans thrown into the bottom of a white concrete monument honoring William Dyess. Despite neglect, the colony remained young and full of promise in the memories of many former settlers.

William Harve Smith and his family were the first to move into the colony during October 1934. In a January 9, 1938, *Arkansas Gazette* Smith shared his feelings about Dyess, which probably reflected the views of many residents throughout the town's history. "There's nothing against a man here. Here, they've given a man a chance when it looked like no one else would. I've got a home and some good land — best in the world. It's the best proposition a poor man ever had."

A New Deal in Dyess

    After going back in time some final observations about Dyess are positive and others negative. Author Paul Keith Conkin studied New Deal communities and admits that the finale to their story was written in an age far "removed from the insecurity and intellectual volatility of the Depression. It was written in terms of reaction. The idealism and reforming zeal of the architects of the communities and the new society were repudiated. The revolution was over. Traditional gods once again possessed men's minds and claimed their loyalties. Most people soon forgot that Dyess, Arkansas . . . had been part of a large social experiment."[123] Desperate farmers who sought a place there believed the colony offered life in a promised land, and many say it did. But some considered it a land of broken promises. Tensions generated by outsiders and insiders prevented Dyess from becoming what it set out to be. The causes for its problems can be laid at several doorsteps. Conservative national politicians attacked the concept as un-American and worked to severely hamper support for such projects. Arkansas officials put Dyess in the middle of a turf war. Powerful planters whose empires surrounded the colony undercut it, complaining that it drove up wages and benefits. That it did, thus threatening the slavery by another name that many imposed on their tenant farmers and laborers. Some federally appointed administrators and managers in the colony attempted to control farm operations down to tiny details. Often they communicated their instructions with an insufferable arrogance. A notoriously independent lot, farmers reacted as one might expect, with contempt and complaints. So it all passed by like a caravan of wagons with squeaking wheels. Though Dyess Colony came and went that way, it still is remembered fondly by many men and women who lived there. And what they endured survives in songs sung by a man dressed in black.

---

    [123] Paul Keith Conkin, *Tomorrow A New World. The New Deal Community Program* (Ithaca, N. Y.: Forgotten Books for Cornell University Press, 2012), 233.

COLONY CREDO

Published in the October 23, 1936, *Colony Herald:*

> *Richer lands for every farm. A home of beauty, comfort, and convenience for every family. A "Blue Ribbon Farm Family" in every home. Modern equipment for every worker. Club work and vocational training for every child. Every crop from purebred seed; every animal from a purebred sire; support of farm ordinations [sic]; farm and home agents, and cooperative marketing for every family. "Equality for agriculture" in every form of legislation, taxation, and education. Full gardens, corncribs, smokehouses, and feed barns to insure a good living at home plus at least two money crops [each year] and important income from cows, hogs, or hens as the business policy of every farmer. A love of country, community, church, and school – along with recreation through books and music, and art – to enrich and ennoble life for every individual.*

A comfortable home for every family was part of the Colony Credo. Colony developers attempted to provide a variety of house sizes and floor plans to meet needs. On opposite page, clockwise from top left, homes of Byron B. Wilson, Emmett Yancey, Richard Hurst, Clarence Edmonston, Willie Bailey, Thomas Green Mitchell, Dewey Owen, and Hershal Henson. Family Collections

APPENDIX A
Glossary of New Deal Agencies and Programs

Agricultural Adjustment Act (AAA): a federal law enacted in 1933 that reduced agricultural production by paying farmers subsidies not to plant on part of their land, thus reducing supply and increasing demand.

Arkansas Emergency Relief Administration (AERA): FERA's state division, headed by W. R. Dyess of Mississippi County, Arkansas.

Arkansas Rural Rehabilitation Corporation (ARRC): established as a separate corporation under AERA, with initial responsibilities for Colonization Project Number One (Dyess Colony).

Civilian Conservation Corps (CCC): one of Roosevelt's first New Deal programs, the CCC was a public work relief program to provide employment for unmarried men.

Dyess Colony Cooperative Association (DCCA): established in 1936 to represent all colony cooperative enterprises, including the blacksmith shop, café, laundry, store, hospital, furniture factory, cannery, service station, shoe shop, cotton gin, garage, sorghum mill, and a printing business.

Dyess Colony, Incorporated (DCI): an official ownership/management entity established in 1936 for the colony, although the board of directors remained primarily federal administrators.

Dyess Farms, Incorporated (DFI): became owner of the project in June 1940, following a report from FSA on changes that needed to be made for the colony to remain viable.

Dyess Rural habilitation Corporation (DRRC): established in 1937 as successor to DCI.

Farm Credit Administration (FCA): established in 1933 to help farmers refinance mortgages over a longer time at below-market interest rates.

Farm Security Administration (FSA): a New Deal effort to combat rural poverty, established in 1937 as successor to the Resettlement Administration.

Farmers Home Administration (FHA): established in 1946 to extend credit for agricultural and rural development. It replaced FSA.

Appendix A

Federal Emergency Relief Administration (FERA): one of Roosevelt's first New Deal programs, it replaced Hoover's Emergency Relief Administration (ERA) and gave grants and loans to states to operate relief programs. WPA replaced it in 1935.

Resettlement Administration (RA): set up in 1935 to relocate out-of-work families (urban and rural) to communities planned by the federal government.

Rural Rehabilitation (RR): a division of FERA which enabled poor families on relief to purchase land and capital goods using long-term loans.

Southern Tenant Farmers Union (STFU): integrated farm laborers union established in 1934 in response to Agricultural Adjustment Act abuses.

Work Projects Administration (WPA): successor organization to Works Progress Administration in 1939, retaining the same acronym.

Works Progress Administration (WPA): largest of the New Deal programs, WPA employed millions of people through public works projects. In 1939 it was renamed Work Projects Administration.

APPENDIX B
Questionnaire for Prospective Rural Colonists

A. <u>Husband</u>

1. Name _____
         (Last)   (Middle)   (First)

2. Age_____Height_____Weight_____

3. Place of Birth _____City_____Small Town_____Farm_____

4. Nationality _____Nationality of Father_____Mother_____

5. Profession, trade or usual occupation of husband _____

6. Religion _____Religious Interest: (a)Very much____ (b) Some____ (c) None____

7. Health (check one) Excellent_____Good_____Average_____Poor_____

8. Education _____Where _____ No. of Years _____

9. References:
           Doctor_____
           Grocer_____
           Minister _____

10. Farm Experience _____Years _____; Gardening _____Years _____

11. What creative interest in music, arts, crafts, theatricals, etc. _____
    _____

12. Non-creative interests – athletic, reading, collecting, etc. _____
    _____

13. Special skills, or training? _____
    _____

14. Physical defects, if any? _____

15. Employers: Last three for whom he worked a month or more, giving types of work
        (1) _____
        (2) _____
        (3) _____

APPENDIX B

B. Wife

16. Name _____
          (Last)   (Middle)   (First)
17. Age_____ Height _____ Weight _____

18. Place of Birth _____City _____ Small Town _____ Farm _____

19. Nationality _____Nationality of Father _____ Mother _____

20. Profession, trade or usual occupation of husband_____

21. Religion _____Religious Interest: (a)Very much ____ (b) Some ____ (c) None ____

22. Health (check one) Excellent _____Good _____ Average _____ Poor _____

23. Education _____Where _____No. of Years _____

24. References: Doctor_____
                 Grocer_____
                 Minister_____

25. Farm Experience _____Years _____; Gardening _____ Years _____

26. What creative interest in music, arts, crafts, theatricals, etc. _____
    _____

27. Non-creative interests – athletic, reading, collecting, etc. _____
    _____

28. Special skills, or training? _____
    _____

29. Physical defects, if any? _____

30. Employers: Last three for whom she worked a month or more, giving types of work
    (1) _____
    (2) _____
    (3) _____

Source: Family Selection Section, Resettlement Administration, April 25, 1936. Record Group 96, Farmers Home Administration, AK-80, National Archives. Copy in Dyess Colony Archives, Arkansas State University.

Appendix C
Case Worker Analysis
of a Prospective Colonist

COUNTY _____ CASE WORKER _____ DATE _____

NAME OF FAMILY

_____
    Man's Last   Man's Middle   Man's First

NAME OF WIFE _____ MAIDEN NAME _____

ADDRESS _____ CASE NUMBER_____

1. Is there any friction in family, or do their personal interests conflict?
2. Who is the dominant member of the family?
3. Is there any evidence of hereditary weakness, mental or physical?
4. Is there any health problem in the family?
5. Are there any personality problems in the family? (Moods, temperamental peculiarities)
6. Does the family get along harmoniously with the neighbors?
7. Has any member of the family any bad habits? Drunkenness, loose morals, inconsideration of the sensibilities of others, etc?
8. Would you consider this family of high, medium or low intelligence?
9. In what community activities do the members of the family participate?
10. Is there any evidence of dishonesty in the family? Do the children indulge in petty thievery?
11. Have they been known to the probation or police departments?
12. Has the family tried to meet its obligations?
13. Have they cooperated with the case workers?
14. Have they made the best use of their assets and income? Have they been resourceful?
15. Have they tried to help themselves? Have they initiative? (Give facts)
16. What was their social and economic status before the Depression?

APPENDIX C

17. What is the religious affiliation of the family?
18. Is the feeling extreme or emotional?
19. Is the family extreme or emotional in its political beliefs?
20. Is the man industrious? Is he a hard worker?
21. Does he try his best to provide for his family?
22. Is he interested in his family? Does he care for them to the best of his ability? Does he watch for their welfare?
23. Is the woman a good housekeeper? Does she keep the house clean? Is she a good cook?
24. Does she take care of the children properly?
25. Is she an efficient manager?
26. Considering the limitations of the family's resources, is the home fitted up with taste and dignity? Are they interested in their home?
27. Do the older children have a feeling of responsibility toward the family?
28. Have any of the members of the family special talents, especially in the arts, crafts, music, theatricals, etc?
29. What was the credit rating of the family during the time it had an income?
30. Did it meet its obligations? What do former creditors say?
31. What does the school report as to the intelligence of the children?
32. As to the cooperation and interest of the family in the children's welfare?
33. Has the family maintained a garden? How many years?
34. How many quarts of vegetables put up this year? Fruits? Meats?
35. What canning methods do they prefer?
36. Does the woman, or other member of the family, make any of the family clothes?
37. Have they made or painted any of their furniture, made draperies, rugs, etc?
38. How well do you know the family, and how complete have been your contacts with them?
39. Other comments:

Source: Family Selection Section, Resettlement Administration, April 25, 1936. Record Group 96, Farmers Home Administration, AK-80, National Archives. Copy in Dyess Colony Archives, Arkansas State University.

Some Dyess children, clockwise from top left, Herbert Roberts, Ann Roberts, Elbert Geater, Dewayne Roberts, Marlin Joe Roberts. Courtesy of Ann Roberts Blue

APPENDIX D
Applicant Interview

This interview took place between a Dyess Colony selections specialist and a prospective colonist at the applicant's home located 14 miles from Russellville, Arkansas. It has been edited.

Specialist: "I am Mr. S. from the regional office in Little Rock. I have come to talk with you about your application for a home in Dyess colony. Mr. and Mrs. Doe, I believe through your rural supervisor, Mr. X, you placed your application with the regional resettlement office to purchase a farm at Dyess Colony."

Mr. Doe: "Yes, we did some time ago, but had decided it had been forgotten until Mr. X called us into his office a few days ago. He told us some about that place, but said some man would be out here to see us in a few days, would tell us about it and check up on us."

Specialist: "Well, I suppose I am the man he referred to, but my primary purpose is not to check up on you, but to discuss your plans with you and give you what information I can about the project so that you will be in a position to decide more wisely just what you wish to do. I have not been sent out here by the regional office to tell you and Mrs. Doe what you can or cannot do. You can do that by yourself. Every family has a head of family. Who is the boss here? Mrs. Doe, probably the women boss us, don't they? And as such they must be responsible for all the moves and decisions of the family. Mrs. Doe, I have been sent to assist you and your supervisor in preparing your application. After your application has been completed it will be sent to the regional office for consideration. I am here to help you in any way possible, and I want you to ask questions. In fact, I will be disappointed if you do not. So feel free to ask any kind of question any time it occurs to any of you. Are there grown children at home? I'd be glad to have them discuss the matter with us for I am sure they have questions to ask."

Mrs. Doe: "No. All are in school, but the oldest boy. He is working for a neighbor today. He went to school until last year, but times got so hard, and we had so little, that he decided to drop out and work at whatever he could find to do. The oldest girl stayed at home today to help me wash and scrub, but she is out just now."

Specialist: "I was born and reared on a farm. Mother and Daddy had a large family, and I realize what a job it is to feed, clothe, and keep all the children interested in school. How are the children getting along?"

Mrs. Doe: "Just fine. You know they go to town to school now. At first we were against taking the children into town, but we are glad of it now. If we still had to send them here, none of the children would go to high school. The teacher tells us all are doing well."

A New Deal in Dyess

Specialist: "Mr. Doe, how long have you and the family considered going to Dyess Colony?"

Mr. Doe: "It has been only two or three weeks since our supervisor was out here and talked to us about it. We asked him several months ago to help us get a government farm or help us rent where we could have bottom land. For the past 10 years we have been on the move until two years ago when we came back to the old home site and decided to stay here until we could find a place where we could make a living, and have a permanent home."

Specialist: "Mr. Doe, do you think your family really wants to move to Dyess?"

Mr. Doe: "Well, we could hardly know unless we could see the place. Will they let us go over there before we decide?"

Specialist: "Both you and Mrs. Doe are expected to go. They prefer all grown children who intend to move to the colony go with you because if the entire family is not satisfied, you will not stay and be successful. You know, Mr. Doe, a move like this is important, [and] it should not be done hastily or without thought and planning by the whole family. I am sure you would not buy a farm from some individual or land company and move your family into a strange place unless you had seen the place and thought you knew thoroughly what you were doing. If this move is worth making you will live there for a lifetime. If you do not make it a permanent home and should leave in a year or two no doubt you will leave in a worse condition than when you went there. It will take time to develop a home and establish yourself in a new location."

Mr. Doe: "You are right. Several years ago I owned my own team, tools, and other stock. I left and went to Oklahoma where we had a drought. I lost my team, and since that time I have moved every year until I came back here. We have done lots of hard work on several places, then would have to leave it. I know we could make a living if we had good land and could live on the same place year after year. What kind of land is that at Dyess?"

Specialist: "It is cut over land. They call it 'buckshot.' About 16,500 acres of land were bought from a large lumber company. It has been surveyed into 20, 30, and 40-acre farms. Roads have been laid out, and drainage ditches constructed across the track leading into the main drainage ditches and the Tyronza River. There is one thing you can be certain of, and that is the richest of soil and a good house in which to live."

Mr. Doe: "Does that country flood every time the Mississippi gets up?"

Specialist: "As I understand that land was partially covered with water in 1927. It was not overflow water from the river, but it accumulated from excess rainfall. At that time this land was not cleared up and drained. Since then the Tyronza River has been cleaned out, ditches constructed, and I feel sure there is very little danger of future

APPENDIX D

floods. I don't believe the Government would spend three million dollars developing this project and move 500 families in there unless it was sure there is no serious danger of future floods."

Mr. Doe: "How large is that river?"

Specialist: "It is a very small, narrow river. Should you cross it then cross one of the largest drainage ditches you would hardly know the difference. It is nothing like the Arkansas River you are acquainted with."

Mr. Doe: "Well, tell me are there any fish in that river?"

Specialist: "I suppose so, but whether they are in large number I do not know. Do you ever fish and hunt?"

Mr. Doe: "No we don't. Usually we will all go about twice a year, and camp out one or two nights. Sometimes another family goes with us. I hunt a little during the fall and winter. The boy has hunted more this year than usual. We make some money at it. Last year I made enough to buy the family's clothes."

Specialist: "Mrs. Doe, don't you like to fish?"

Mrs. Doe: "Yes, better than any of us."

Specialist: "I think it is a very fine sport, and I would like to fish more than I do."

Mr. Doe: "I know you are going to know everything about us before we are sold one of those farms, and I want to tell you one thing before the children come in."

Specialist: "Feel free to discuss anything with me."

Mr. Doe: "When I left here and went to Oklahoma, and settled for a crop, it was not long until another man from this community moved to the same place in Oklahoma and settled on a farm near me. He had left Arkansas because he was in trouble. I had known him all my life. He was a desperate character and had been in the pen before. He had been accused of killing one man, and was not convicted for it. He had come to my house several times and wanted to get me to help him to do work on a contract. I wouldn't, and he got mad about it. I asked him to stay away from my house. He came back later and was drunk. He had previously threatened to kill me, or do my family harm, or burn my barn. I was afraid he would. He began to abuse the children out in the yard, and I had to protect myself and family. I killed him. I was brought to trial, and character witnesses were carried from here out there because they had known both of us all our lives. I was not convicted, and the jury decided the case in five minutes. If this will eliminate me I would like for you to drop my application now. If it doesn't, I'd be glad for you to talk to the leading citizens around here. I have never talked it

A New Deal in Dyess

with but three people. They are ...., ..., and .... Everyone knows about it and can tell you the facts of the case."

Specialist: (The worker had previous knowledge of this past trouble in Oklahoma and was convinced that Mr. Doe was justified.) "Mr. Doe, self preservation is the first law of nature, and in all countries a man has the right and is obligated to protect himself and family. Our American system says that a man shall be tried by an impartial jury. You have been. So you have satisfied the law. You have returned to the home community where both you and the other man were known. You have lived here two years, and your home has accepted you. They are the best judges in the world. Any of the people would be glad to have you as a neighbor, and so far as we are concerned that satisfies us. I do think there is one thing you and your family should consider. Here all is well, and your home people know the facts in the case and have never doubted you and your good family, but when you go into a strange community among strange people would it embarrass you for those people to know about this past? Would it cause you any trouble? We will never tell it, and possibly it will never be mentioned again for it is none of our business, but other of your friends from this community are going to move to Dyess and will in all probability tell it some time. If you think that will cause you no future trouble I see no reason why it should trouble you now."

Mr. Doe: "I am not afraid but what I can go anywhere and get along with any average person, and I'm not afraid but what my neighbors will like me."

Specialist: "If you feel that way about it, I have no fears. Did you ever have any trouble with this man before you left here?"

Mr. Doe: "No. I was like everybody else. I had as little to do with him as possible. That is the only other time any member of either family was ever in court or had any trouble with anyone."

Mrs. Doe: "We have never had much sickness and would not want to go to an unhealthy place, and we have heard so much about Dyess that we wonder if we would have good health."

Specialist: "That is a question no one can answer, for you might move to any place and have ill health. That depends on the physical condition of the family and the care they give themselves."

Mr. Doe: "What kind of water do we have to drink?"

Specialist: "It is what you call hard water. It won't do the laundry well, and of course that disappoints the women, for they usually have to do the family wash. The colony does not have a general water system as you may have heard, with all the water coming from one deep well. Each house has its own well supplied from a driven pump. These pumps are driven on an average from 40 to 80 feet deep, and that depth will make the water safe. You will have the same water to drink that these people in the Delta section have always had. When you ask about health you were thinking about the water and mosquitoes. You will have lots of mosquitoes, and unless you protect yourself you will have malaria, which is contracted from mosquitoes [and] not water as we sometime believe. Your house will be well screened, which will offer you protection from this danger. That section is well drained, and year after year more of that land should be put into cultivation. This will tend to destroy breeding places which will gradually cut down the mosquito danger. I am sure there will always be mosquitoes there, and unless people protect themselves they will have malaria. I think the health your family has there depends largely on the precautions you take. You will need to take the same precautions there as here."

Mr. Doe: "We have lived in black land where there are as many mosquitoes as anywhere, and we were healthy enough. What kind of farming do they do?"

Specialist: "That is primarily a cotton country, and cotton will probably be your cash crop. The plan is to diversify crops and operate on a live-at-home basis. Some people will have their cows, sell milk, butter, and cream. Some will have a large flock of chickens. Nearly any kind of vegetables will grow there, and it is planned that you will devote more attention to growing food crops than most families have in the past. The colony provides a large canning plant which will put out 42,000 cans per day, and it is hoped that each family can have its own pressure cooker and during the summer months be able to save enough vegetables to last until spring gardens come in again."

Mrs. Doe: "We have our pressure cooker, and it has been a great help to us. We have had lots of food to waste because we could not get cans. The R.R. promised [us] plenty of cans, but they ran out of money or something, and we did not get near all we needed."

Specialist: "The whole project is worked out on a cooperative basis. For example, it is hoped all will grow the same kind of cotton and have it ginned at the colony gin. If this is done officials of the colony can guarantee pure seed and sell them on the market as planting seed instead of 'mill run,' which would give you about $60.00 to $80.00 per ton for your extra seed instead of $10.00 to $30.00 you would get if each grew a different kind of cotton (and) sold it as 'mill run.' It is planned to experiment with other market crops. For example, it might be hard to find a market if only one or two families planted a small patch of early corn for roasting ears, but if all the families or a large part of them should plant a small plot, all of them would give a large enough bulk to truck to Memphis or St. Louis. They would not want a family to plant all its land in one crop like this, for if this crop failed the family would be in the same condition we have been in the past when a flood or drought destroyed our cotton crop. After this early crop has been harvested the land can be planted in another crop. Should this crop fail, no family has been hurt much for you would not have 'all your eggs in one basket.' They plan the same things with other crops and farm products. You know when a whole community goes together it can do far more than an individual."

A New Deal in Dyess

Mr. Doe: "Well that sounds good to me, and I'd be willing to work on a plan like that."

Specialist: "The administration plans cooperative organizations. For example, colonists have the same breed of milk cows. If they decided to do so then by a vote they would decide on the breed. Those are only a few examples."

Mr. Doe: "Since we do not know what kind of chickens, hogs, cows, and crops the colonists will have, would we take what we have or sell them here in case we should go?"

Specialist: "I would advise you to move anything that is worth moving. You should take everything which you found useful here. In your case, I would sell all my stock, except the work stock and good milk cows and hogs for my next year's meat, because you will have to pay transportation, buy feed for them, and this will soon amount to more than their value."

Mr. Doe: "My mules won't be large enough for that work, will they? They weigh only 800 lbs."

Specialist: "I doubt if they will be large enough, but mules are very high, and I would move them. They will exchange [them], and you pay the difference."

Mr. Doe: "Most of my plain tools are not suited for that land. What would you do with them?"

Specialist: "Since most of your tools were furnished by the R.R., I would advise you to take the matter up with Mr. X. He will possibly take them back and give you credit. If not, take them along, and then the colony officials will trade with you. If you can get what they are worth in money, sell them to save transportation. You should be a better judge of what to do with the equipment you cannot use than we are. Always do whatever will net you the most."

Mrs. Doe: "Would you move your empty fruit jars and cans?"

Specialist: "I surely would for they will be hard to secure there as here and will cost as much or more money."

Mr. Doe: "Will we get there in time to plant a crop this year?"

Specialist: "You will not get in a full crop, but the plan is to move each family early enough to plant a partial crop. You must keep this in mind – you can plant crops there much later than you can here, and you will not have as many acres to produce more. As a rule that land will make a bale of cotton per acre, and 25, 50, 40 bushels of corn. It will produce together crops in about the same proportion."

APPENDIX D

Mrs. Doe: "Here we make about a bale to every 4 or 5 acres, about 5 to 10 bushels of corn per acre, and sometimes nothing but cut feed. Is a 20-acre plot of land all we can get?"

Specialist: "Most of the farms have about 20 acres in them. A few have 30 and 40 acres, but these will be assigned to the large families that have plenty of labor to help work it. You must remember [that] 20 acres of this land in terms of production is equal to 40 or 60 acres of the land you have cultivated here. According to the size of your family, I would not expect, if I were you, more than 20 acres. That acreage may seem small, but the project is planned to give families an income sufficient to maintain a decent standard of living. It is not designed for big farming or big machinery—just a place where you and your family can make a living and be secure year after year. The land and the improvements on it will cost you from 80 to 100 dollars an acre to be paid for over a period of 20 to 40 years. Every acre you add to your tracts adds that much to your total debt, and the proportionate amount to your annual debt. If I were you I had rather have a small tract on which I could live, pay taxes, and make my annual payments than have a larger tract and have such a large annual payment. I might not be able to meet it and eventually lose it."

Mr. Doe: "I expect that is right. I had not thought of it in that way."

Specialist: "The first year you go there on a lease or rental basis. At the end of the first year you will be given an option or opportunity to buy it. If you and your family are still satisfied, and you feel you have not made a mistake, then you will enter into a sales agreement at which time you will be told what your annual payments will be. When you have paid for the place you will be given a deed just as you would here in the county should you buy land from an individual or a bank. The place will not be deeded to you until paid for because a deed would make it yours. When you enter into the sales agreement the title will be transferred to you, and then you will be assessed, and pay tax the next year. That would make it about three years at least before you actually begin paying taxes."

Mr. Doe: "Why can't the officials tell us exactly how much the land will cost, and how much we will have to pay each year?"

Specialist: "Because the construction is not completed, all the houses are not built, and another year will be needed to do this. Then, too, it is planned to give work to the colonists in completing the construction, rather than outside people. That cannot be done if families are not allowed to move until the project is completed."

Mr. Doe: "Well, that is a good thing. I'm not afraid the government will not be fair with us, but of course, I would like to know."

Specialist: "Mr. and Mrs. Doe, the purpose of that whole project is to make it possible for 400 or 500 families to make an independent living by hard work. I don't think a penny will be given to you, and I don't think you will be cheated out of a penny. It will cost you what it has cost the Government when completed. No more and no less, as the Government doesn't expect to make money."

Mr. Doe: "Will we have to help clear that land?"

Specialist: "I see in the newspaper that you will receive help in clearing 10 acres. We have not been told that officially, and I would expect to clear my own land. Then if you get help it will be a happy surprise. If you are helped it will be charged against you, and if you could clear your own land your debt would be less."

Mr. Doe: "Will we be given so much work to do each month?"

Specialist: "I do not know. As it is planned you will possibly give some work to pay for your furnish, but you and the family will be too busy at home to be away working on the project."

Mr. Doe: "We have been told that we would be paid for working on our own farms."

Specialist: "That sounds too good to me to be true, and I doubt seriously if it is. This is not a relief project, and the purpose of it is to take your families off of relief, and I have been told that you will not be given a thing. Suppose you were going to be given some work. It would not last long for all will be done, and the work outside of the farm will be as scarce as it is here. So when considering this move I would not let the possibility of getting a little work for a while influence my decision. You are not assured of that, and if you were it would not be for any length of time. The thing your family should decide as far as the financial side of this move is concerned is, 'Can my family pay taxes and make a small annual payment on 20 acres of land after we have had a little help for one year?' That is the thing you will have to do."

Source: Family Selection Section, Resettlement Administration, April 25, 1936. Record Group 96, Farmers Home Administration, AK-80, National Archives. Copy in Dyess Colony Archives, Arkansas State University.

APPENDIX E
Colony Officials and Roster of Colonists by County of Origin as of May 1, 1936

BOARD OF DIRECTORS, DYESS COLONY, INC.
Floyd Sharp
H.C. Baker
R.C. Lamerick

ADMINISTRATIVE PERSONNEL
E. S. Dudley, Resident Manager
J. E. Terry, Assistant Administrator, Agriculture
Cone Murphy, Superintendent of Construction
Fern Salyers, Home Economist
J. G. Womick, Director of Education
H. C. Davidson, Colony Engineer
Clark C. Tucker, Senior Selection Supervisor
R. O. Rainwater, Disbursing Officer
J. S. Clinton, Colony Mercantile Manager
R. C. Lancaster, Acting Warehouse Foreman
Dr. L. L. Hubener, Chief Medical Officer
Dr. J. H. Hamner, Colony Physician
Miss Shaw, Director of Recreation

COLONISTS (Note: There are 428 colonist families included on this May 1, 1936 roster. By July 17, 1936, the number of families had increased to 470, and by fall 1936, the Dyess Colony reached its peak of 487 colonist families. *Source: Dyess Colony Herald*.)

Arkansas (11):
Ed Ward, Seab Staten, Silas Edington, Richard Gibbs, L.W. Brantley, King D. Craig, Guy B. Wheatley, Emmett Yancey, Henry Hastings, John Ramer, Roy A. Goacher.

Ashley (16):
J.E. Echols, G.L. Morres, W.T. Lovett, B.M. Woods, J.T. Locke, James B. McCann, Daniel E. Wilson, John E. Chambers, Robert R. Holland, Loyd Ward, Austin V. Wasson, Ed A. Jones, Garland Scott, William M. Dochala, Dewey McCone, John Christmas.

Benton (1):
Alford H. Campbell.

Boone (2):
Herbert B. Seitz, Harold E. Mulford.

A New Deal in Dyess

Bradley (8):
Thomas Thrasher, J.G. Bratton, M.W. Jackson, Warner Hargraves, Clifton D. Cunningham, John W. Williams, Earnest T. Adair, William O. House.

Calhoun (10):
Claude Seymore, William L. Jacobs, Millard M. Garrett, Sam Massey, Chester Bowen, William Fritts, Carlton Ethridge, Ethridge Waldrip, Chester E. Pierce, James L. Jacobs.

Clark (5):
Oscar Ward, Carl Schmalhorst, J.C. Wingfield, R.O. Harris, W.N. Carter.

Clay (9):
Riley Kazee, Oscar E. Smith, B.R. Halford, Floyd Slayton, Harold Cline, Harry Myers, Edger Pulliam, Arnold Cox, Homer D. Donaldson.

Cleburne (2):
Franklin Huff, Orville F. Farmer.

Cleveland (7):
Sidney Doster, Ray Cash, Buster Dykes, Clyde Brazelton, Roma L. Paden, Carl Word, Alfred Sims.

Columbia (6):
Sammie J. Horton, W.E. Marlar, W.D. Owens, Tom Reading, Dudley C. Thrailkill, Prentis A. Elmore.

Conway (11):
Roy L. Martin, Aubrey L. Crawford, Charles Yarbrough, Martin Ward, Bud Wallace, Thomas A. Stover, Arlin D. Edwards, A.L. Milam, Jack McCraven, H.H. Crawford, Albert J. Spier.

Crawford (4):
Burl J. Cook, Joe F. Southard, Melvin L. Stanford, P.A. Harker.

Craighead (10):
Emory Hall, Millard Simmons, C.T. Knight, Clinton Bowers, Earl Erwin, Robert F. Fortner, Allen C. White, Homer C. Wooten, Solomon Bradford, George Higgenbotham.

Crittenden (3):
R.L. Powell, Elmer H. Craig, Bill Minor.

Cross (7):
Garnett L. Beggs, Sam Hanks, Curtis Jones, Henry C. Brewer, Arthur J. C. Thorne, Baxter Walden, Hubert Copeland.

APPENDIX E

Dallas (10):
Olgar A. Kern, Oscar Marshall, Lloyd Cox, William H. Phillips, William P. Scrimshire, Davis Bradley, H.L. Holloway, Ira M. Looney, Joe D. Livingston, Walter Cox.

Desha (5):
Daniel Rice, Bert Handley, Cyrus Fudge, Thomas J. Reeves, E.L. Johnson.

Drew (3):
Joe Lankford, G.M. Bene, Burton C. Wampler.

Faulkner (14):
Tom Braden, Everett Morgan, Otis Barton, E.J. Hatzell, Dewey C. Fielder, Carl L. Padgett, Jasper J. Allen, Sidney L. Whitaker, Tom J. Ballew, Lester Spear, Fred W. Thomas, Homer C. Chaney, William L. Henry, Homer B. Black.

Franklin (4):
Lemuel S. Satterfield, Alva Taff, Ulysses L. Malott, Alford W. Shores.

Fulton (4):
Walter L. Matlock, William Woods, Bert L. Mathis, Oscar L. Matlock.

Garland (1):
William T. Tapp.

Grant (12):
H.H. Woodson, O.B. Gean, A.H. Whitlock, A.W. Ledford, Frank J. Rutherford, William Rankin, Bryant Smith, Ross Roberts, William Kinday, Mose G. Ashcroft, Luther M. White, Bazzil J. Wilson.

Greene (2):
Walter L. Berry, E.B. Eubanks.

Hempstead (3):
C.R. Rosenbaum, R.L. Williams, David Hunter.

Hot Springs (1):
Pete Orr.

Independence (10):
Walter Wyscaver, J.L. Hancock, B.J. Kent, Cecil R. Mooney, Arthur G. Nichols, Jacob Balch, Lyle W. Johnson, Orvill Alexander, Warren G. Wyscaver, Cecil Watkins.

Jackson (4):
R.L. Hurst, D.L. Staley, Homer Williams, Troy Stansbury.

A New Deal in Dyess

Jefferson (4):
A.B. Craig, O.L. Morgan, Henry Woolery, Almos Overton.

Johnson (8):
A.L. King, George Linton, A.M. Shelton, Claude Slack, Frank D. Tyler, Marvin J. Harmon, Herman C. Pierson, Hugh Bean.

Lafayette (7):
Garland A. Page, L.M. Keen, Richard H. Wooten, Lewis Harrison, Wilson Z. Cheatham, T.T. Gordey, Ardis T. Allen.

Lawrence (9):
Max Tillman, Hershal O. Henson, Quint J. Dent, Lee S. Stow, William R. Metcalf, Bill J. King, Elmer D. Mooney, B.G. Price, Etsel Hathcock.

Lee (7):
Fred A. Alderman, Buff Roberts, David M. Howard, Herman Tiner, Doxie W. Reeves, Luther L. Long, Willie Bailey.

Lincoln (13):
Bester Prysock, Tommie E. Thompson, Christopher E. Bennett, Warren Bramlett, Joe V. Wilson, John B. Newsom, John J. Butler, Garner M. Phillips, Jesse A. Rhoads, Johnnie Culpepper, Elmer C. Rogers, Ernest C. Hunter, Roscoe Phillips.

Little River (1):
Oscar M. Mitts.

Logan (3):
Robert Z. Beck, R.E. Roberts, Veasey Haller.

Lonoke (10):
John Thompson, Ralph E. Farish, Earl Waters, Alonzo A. Griffis, John A. Counts, William N. Bell, Edward R. Dodd, Lee McChesney, Theo Gouchenour, Charles Cox.

Marion (3):
Don Smith, Ray J. Noe, Jack S. Conley.

Miller 11):
Claude Creamer, C.V. Coe, Clarence L. East, Calvin Ramsey, S.D. Sams, Wylie W. Bayless, L.M. Doss, Clarence Belk, John H. Dalrymple, Claude W. Wharton, Ben H. Hendrix.

Mississippi (18):
H.J. Richardson, E.D. Pickens, Thomas Dickerson, Vernon Henry, Ernest C. Pickens, Clarence W. Edmonston, Q.A. Bullard, Doss Halle, W.L. Williams, E.A. Pickens, R.M.

APPENDIX E

White, J.F. Russell, M.P. Monday, Roy Fennel, W.H. Smith, Ivan Butler, W.T. Dally, J.H. Burfield.

Monroe (6):
Milwee Bryant, Hugh Griffith, Jesse L. Pilkington, Marshall W. Plumlee, Dawson C. Deaton, E.M. Swafford.

Montgomery (1):
Clifford W. Pate.

Nevada (7):
Lloyd Sampson, Harvey Hamm, Howard Fincher, Terrell C. Brantley, Byron B. Bevill, Willie Hodnott, Herman Kimbrough.

Newton (1):
Roy C. Carlton.

Ouachita (6):
Curtis White, Fred Weaver, Christian Seiler, Homer P. Snearly, Floyd J. Mosley, Barney Nutt.

Phillips (9):
Robert A. Foley, Newell E. Strawn, John A. Wallace, Clarence Smith, Sim C. Newberry, Tom Hale, Homer C. McDaniel, Fred D. Humphreys, Charles F. Culwell.

Pike (3):
Dewey L. Cox, Everett Sutton, Byron B. Wilson.

Poinsett (5):
Dewey Smothers, Mack Howard, Otis Minton, Cecil B. Maddox, Sherman H. Wallace.

Pope (6):
Ray Shepherd, Louis Sipes, Garland Chilton, Alex Renfro, Nargus Kraus, Cecil Jennings.

Prairie (6):
C.L. Brooks, Edwin Bowls, Slather Jones, Roland Hill, Clarence Dewey, Henry Best.

Pulaski (8):
Leo Allen, R.N. Hovater, Ed Knight, Ernest Norton, Guy Gilliam, Oille Wagner, Paul R. Anderson, Russell Wilkerson.

A New Deal in Dyess

Randolph (11):
Lee Litherland, Granville Anderson, Jonas L. Myers, Dave Edington, Luther V. Clay, Thomas Cypert, Bob Haynes, Major D. Moore, Charlie Hill, Wm. F. Mitchell, Walter Hooker.

Saline (5):
Wilmer G. Young, Herbert Humble, Dale Burris, Sidney A. Chrestman, S.B. Funk.

Scott (2):
Walter M. Johnson, Lewis Slaughter.

Searcy (2):
Noah Passmore, John Bagby.

Sebastian (5):
R.S. Ringenberg, Kenneth I. Smith, W.J. Gardner, Joe Byers, Granville P. Hardwick.

Sevier (10):
Dale R. Simer, Alvin Gore, Zeb L. Shaddix, Major B. Cornelius, Fletcher Alford, M.B. Moore, Ruel H. Gore, William H. Southworth, William H. Brantley, Haywood J. White.

St. Francis (9):
Homer Harlson, Bedford F. Wright, James R. Clifton, Marion Ratton, Ernest Q. Perkins, James D. Darnell, H.G. Summers, Ray Justus, Arthur T. Thomas.

Union (4):
L.W. Roberts, F.A. Owen, John New, V.H. Humphries.

Van Buren (17):
William R. Corhn, Floyd G. Little, Ben Buggar, Ben H. Johnson, Harley Mabrey, Oscar Mabrey, Gene Holland, J. Perry Neal, Jack Stark, William F. Back, Howard S. Coffin, Claude C. Phillips, Harvey O. Martin, Frank Bennett, Odie Holland, Opie Sohn, Theo C. Phillips.

White (11):
F.E. Bishop, J.B. Creager, Henry A. Raney, Frank H. Ashley, Ed A. Ballentine, Arthur Wheatley, L.E. Smith, Jack Kirkland, R.E. Malin, Robert C. Hargett, Everett Williams.

Woodruff (2):
James Humphreys, G.W. Richmond.

Yell (13):
Tim A. Dinnes, Richard A. Dodson, Ray W. Brazier, Foy F. Ferguson, Louie Lakey, James E. Freeman, Odie Howard, Harvey E. McVay, Burlin Furr, C.M. Nicholson, William T. Fryar, Everett F. Grist, Elmer Boon.

APPENDIX F
Homestead Sales Contracts

HOUSE #	PURCHASER	DATE ARRIVED	DATE OF SALE	ORIGINAL CONTRACT	BALANCE	ANNUAL PAYMENT
001	Tom Hale	11/27/1934	07/14/1937	2,996.94	2,996.94	152.90
002	Claud Seymore					
003	Sidney W. Doster	11/09/1934	07/14/1937	2,577.90	2,577.90	131.52
004	W. M. Hodnett		01/07/1938	2,819.47	2,819.47	243.85
006	Tom J. Ballew		07/14/1937	2,870.40	2,870.40	146.45
007	Floyd Slayton	11/09/1934	12/18/1937	3,033.50	3,033.50	154.77
008	Sidney W. Funk	11/09/1934	07/14/1937	2,999.52	2,999.52	153.03
009	R. R. Hill					
010	C. D. Cunningham		02/05/1938	2,894.83	2,894.83	147.69
011	Dewey Owens					
012	Hancock					
013	G. W. Richmond	11/12/1934	12/31/1937	2,526.38	2,526.38	128.89
014	Warner Hargraves		02/05/1938	2,046.11	2,046.11	104.39
015	Homer Williams	02/20/1935	In Advance		76.48	
016	J. H. Burfield	05/24/1935				
018	W. J. Gardner	05/24/1935	07/14/1937	2,662.96	2,662.96	135.86
019	Thomas W. Cypert	11/09/1934	01/12/1938	2,307.20	2,307.20	117.71
020	Roscoe Phillips	11/27/1934				
021	Paul Cypert	11/25/1936	01/12/1938	2,148.98	2,148.98	109.64
022	L. M. Keen	11/27/1934	12/14/1937	2,711.19	2,711.19	138.32
025	D. L. Staley	03/30/1935	02/05/1938	2,584.03	2,584.03	131.84
027	Odie Wagner	06/10/1935	02/05/1938	2,270.90	2,270.90	115.86
028	H. H. Crawford	04/19/1935	07/14/1937	2,382.42	2,256.42	121.55
029	Ray Fennell					
032	W. E. Marler					
032	J. E. Hancock					
033	Floyd Maddox	03/12/1937	06/07/1938	2,653.59	2,653.59	135.38
035	Hugh Bean Jr.					
036	R. L. Powell	11/26/1934	12/18/1937	2,315.79	2,315.79	118.15
040	Oscar Ward	05/05/1935	12/18/1937	1,960.01	1,960.01	100.00
041	Jack Kirkland	05/24/1935	12/18/1937	2,462.23	2,462.23	125.62
042	Sam Jones	01/20/1937	08/30/1938	2,492.23	2,492.23	127.15
043	Solomon Bradford	03/28/1936	06/07/1938	1,995.84	1,995.84	101.83
046	Elmer C. Rogers					
048	Frank Tyler	03/13/1936	12/18/1937	2,787.32	2,787.32	142.21
049	Herbert T. House					
049	Ware					
050	Walter W. Cox	04/11/1936	01/07/1938	1,939.67	1,939.67	98.96
053	John Easley	01/09/1937	01/12/1938	2,447.63	2,447.63	124.88
054	Riley Kazzee		02/05/1938	2,301.60	2,301.60	117.43
055	Everette F. Grist	04/04/1936				
057	E. L. Johnson					
057	Everette Grist		02/05/1938	2,861.96	2,861.90	146.02

A New Deal in Dyess

House #	Purchaser	Date Arrived	Date of Sale	Original Contract	Balance	Annual Payment
059	C. V. Coe	02/13/1935	01/22/1938	2,222.81	2,222.81	113.41
060	George L. Morris					
063	Sidney A. Chrestman		07/14/1937	3,338.33	3,338.33	170.32
064	Leo Allen		02/05/1938	3,186.22	3,186.22	162.56
065	Theo Phillips					
066	Byron B. Bevill		12/31/1937	2,479.79	2,479.79	126.52
068	C. A. Smalhurst					
068	R. O. Harris					
070	John McAfee		08/30/1938	1,955.01	1,955.01	99.74
071	Lender J. Brantley		07/14/1937	2,051.22	2,051.22	104.65
072	Walter D. Hooker		12/31/1937	2,245.64	2,245.64	114.57
073	O. E. Wagner					
074	J. R. Echols	10/26/1934	08/05/1937	2,810.42	2,810.42	143.39
075	O. L. Morgan		02/05/1938	2,711.46	2,711.46	122.51
078	G. P. Hardwick		02/05/1938	2,379.15	2,379.15	129.89
078	Millard Simmons					
079	Oscar Smith					
079	Houston D. Hood		12/31/1937	2,545.90	2,545.90	129.89
080	Charles B. Sornson	05/01/1936	01/07/1938	2,808.16	2,808.16	143.27
081	Howard Fincher		02/02/1938	2,110.45	2,110.45	107.67
084	Harvey Hamm		02/05/1938	2,116.53	2,116.53	107.98
086	Earl Staley		02/05/1938	2,145.88	2,145.88	109.48
087	Sim C. Newberry		02/05/1938	2,687.11	2,687.11	137.97
088	Everett Williams					
095	William H. Phillips		12/31/1937	2,113.13	2,113.13	107.81
099	Ray Justus		08/05/1937	1,987.19	1,979.04	101.39
101	B. L. Prysock		01/07/1938	2,424.27	2,424.27	109.39
102	L. L. Gordey		01/07/1938	2,296.85	2,296.85	117.18
103	Cravens					
104	Harold Holloway					
104	Martin Ward		02/05/1938	2,076.47	2,076.47	105.94
105	Barney Nutt		01/24/1938	2,183.25	2,143.54	111.39
106	Everett Williams					
106	Hugh Woodson					
107	Wingfield					
109	Floyd. A. Owens	03/30/1935	08/05/1937	2,968.45	2,803.64	151.45
110	J. P. Golden	06/06/1936	08/05/1937	2,506.27	2,506.27	127.87
111	C. W. Wharton					
113	Jack Conley	03/20/1936	02/05/1938	1,929.47	1,929.47	98.44
114	W. T. Fryar	04/04/1936	02/05/1938	2,456.27	2,456.27	125.32
117	Carl Word	04/04/1936	02/02/1938	2,179.56	2,179.56	111.20
118	E. M. Emfinger	05/07/1936	12/31/1937	2,489.60	2,489.60	127.02
119	Ray Shepherd	03/27/1936	02/05/1938	2,047.53	2,047.53	104.46
120	Hanley					
120	Hugh Bean	03/13/1936	02/14/1937	2,245.76	2,245.76	114.58
121	J. B. Newsom	04/10/1936	02/05/1938	2,138.77	2,138.77	109.12
122	William F. Mitchel	03/06/1935	01/12/1938	2,596.05	2,596.05	132.45

APPENDIX F

HOUSE #	PURCHASER	DATE ARRIVED	DATE OF SALE	ORIGINAL CONTRACT	BALANCE	ANNUAL PAYMENT
124	James Humphreys		01/22/1938	2,073.36	2,073.36	105.78
126	A. M. Shelton	03/24/1935	07/14/1937	2,156.76	2,062.72	111.04
127	Leo Allen	03/21/1935				
129	Gilbert Deaton					
131	Fletcher Alford	04/06/1935	02/05/1938	2,879.91	2,879.91	146.93
132	Smothers					
132	Oda Bennett	03/22/1937	08/30/1938	2,044.66	2,044.66	104.32
134	Chris Seiler	02/02/1935	01/22/1938	2,218.20	2,218.20	113.17
137	Henry A. Raney					
137A	L. E. Smith	03/08/1935	01/22/1938	2,215.97	2,215.97	113.06
138	John Butler					
142	M. F. Monday					
143	Floyd R. Peters	03/16/1937	08/30/1938	2,696.57	2,696.57	137.58
145	A. Claude Jewell	06/15/1936	08/30/1938	2,467.69	2,467.69	125.90
148	P. A. Harker					
148	John H. Bagby	02/27/1936	07/14/1937	2,902.90	2,902.90	148.10
149	R. Z. Beck					
151	Dick Grooms					
153	John Dalrymple					
154	Edwin C. Baumez		07/14/1937	1,922.65	1,922.65	101.66
155	Ode Howard	03/04/1936	02/05/1938	2,403.51	2,403.51	122.63
156	J. M. Stevens					
157	Forrest E. Bishop	04/19/1935	07/14/1937	2,000.34	2,000.34	102.06
157	Forrest E. Bishop		02/05/1938	430.30	430.50	21.96
158	John B. Creager	04/19/1935	08/05/1937	2,009.93	2,009.93	102.55
159	L. E. Spears	03/17/1936	02/05/1938	2,156.05	2,156.05	110.00
160	Ardis Allen					
161	Thomas B. Smith	04/29/1936	07/26/1938	2,787.33	2,787.33	142.21
165	E. M. Swaffor	02/06/1936	07/26/1937	2,950.22	2,950.22	150.52
166	Wylie W. Bayless	04/18/1935	07/14/1937	2,081.96	2,081.96	106.22
167	Theo Gouchenover					
168	George Linton	03/13/1936	12/14/1937	2,077.81	2,077.81	106.01
170	Clyde Brazelton					
171	Clyde Brazelton	03/05/1936	02/05/1938	3,048.53	3,048.53	155.53
171	J. H. Balch					
172	John W. Ramer	02/27/1936	12/31/1937	2,244.34	2,244.34	114.50
173	C. T. Knight	04/11/1935	12/12/1937	2,107.99	2,107.99	107.55
174	Roy M. White	10/21/1935	12/12/1937	2,819.18	2,819.19	143.83
175	O. B. Gean	03/08/1935	07/05/1937	2,136.42	2,136.42	109.50
175	O. B. Gean		09/14/1938	561.50	561.50	28.65
179	Burlin Furr	03/04/1936	01/12/1938	2,802.33	2,802.33	142.98
181	Bob Haynes	03/10/1936	01/22/1938	2,050.67	2,050.67	104.62
182	Basil R. Basso	05/02/1936	07/14/1937	2,297.55	2,297.55	117.22
183	D. M. Bass	04/18/1935	11/14/1937	2,087.85	2,087.85	106.52
184	Earnest C. Pickins	03/31/1936	01/12/1938	2,849.03	2,849.03	145.36
187	E. D. Pickins	05/13/1935	01/12/1938	2,798.32	2,798.32	142.77
188	A. S. Whitlock	03/03/1935	01/12/1938	2,181.05	2,181.05	111.28
189	W. O. House	04/08/1936	08/05/1937	2,183.90	2,183.90	111.42

A New Deal in Dyess

House #	Purchaser	Date Arrived	Date of Sale	Original Contract	Balance	Annual Payment
191	Henry Lee McVay	03/04/1936	02/05/1938	2,161.06	2,161.06	110.26
192	Jack Bledsoe	03/04/1936	07/14/1937	2,856.89	2,856.89	145.76
193	L. M. Doss					
195	Granville Hardwick	03/16/1936	02/05/1938	2,048.39	2,048.39	104.51
196	Granville Anderson		12/31/1937	2,746.43	2,746.43	140.12
196	Wheatley, Arthur					
197	J. C. Wingfield					
198	Silas Edington	03/19/1935	01/25/1938	2,500.90	2,500.90	127.59
199	R. E. Roberts	05/01/1935	02/05/1938	2,271.71	2,271.71	115.90
200	R. L. Williams	05/24/1935	02/05/1938	2,879.29	2,879.29	146.90
204	Emory Hall	12/14/1934	02/05/1938	3,108.08	3,108.08	158.57
204	Argus King					
205	Sam Hanks					
206	A. V. Wasson	04/22/1936	01/25/1938	2,280.64	2,280.64	116.36
207	John Richardson	03/09/1935	02/05/1938	2,941.96	2,941.96	150.10
210	Henry Hastings	02/11/1936	02/05/1938	3,181.74	3,181.74	162.33
211	John W. Burfield	03/20/1936	07/14/1937	3,333.34	3,333.34	170.06
212	W. H. Brantley	03/20/1936	06/07/1938	2,489.36	2,489.36	127.01
212	Langford					
213	C. B. Eubanks	03/29/1935	08/05/1937	2,833.24	2,833.24	144.55
216	W. Z. Cheatham	04/15/1936	02/05/1938	3,016.80	3,016.80	153.92
216	Wasson					
217	Burley Woods					
218	Herbert H. Humble	03/30/1935	08/05/1937	2,915.76	2,915.76	148.76
219	J. E. Britton					
221	Marshall Plumlee					
221	Roy Fennell	01/30/1935	08/05/1937	2,199.17	2,199.17	112.20
222	Almos Overton	02/06/1935	11/04/1937	2,252.05	2,252.05	114.90
223	Clarence C. Drewry	01/28/1936	07/14/1937	2,346.44	2,346.44	119.71
224	Franklin Huff		02/05/1938	2,159.36	2,159.36	110.17
224	A. L. Milam					
224	J. J. Allan					
225	Cecil Jennings	03/15/1936	02/05/1938	2,256.71	2,256.71	115.14
226	Robert E. Holland	03/11/1936	01/12/1938	2,428.79	2,387.04	123.92
227	Oscar N. Mitts	03/02/1936				
229	W. C. Johnson					
230	W. L. Johnson					
231	Seab R. Bagwell		01/12/1938	2,871.37	2,871.37	146.50
233	Troy S. Stansbury	03/27/1935	02/05/1938	2,259.23	2,143.97	15.26
234	Herman Kimbrough	03/13/1935	08/05/1937	2,097.35	2,097.35	107.00
236	E. J. Hatzell					
238	A. L. Holland	06/03/1936	01/12/1938	2,593.18	2,593.18	132.30
239	Tom Braden	05/24/1935	01/12/1938	2,680.25	2,680.25	136.74
245	Willie Bailey	04/10/1936	07/14/1937	2,094.24	2,074.94	106.87
246	Quinoet A. Bullard	02/14/1936	07/14/1937	2,705.77	2,590.14	138.05
248	Vernon Henry	09/15/1935	08/05/1938	2,589.77	2,589.77	132.11
249	M. M. Garrett					

APPENDIX F

HOUSE #	PURCHASER	DATE ARRIVED	DATE OF SALE	ORIGINAL CONTRACT	BALANCE	ANNUAL PAYMENT
250	Wilmer G. Young		02/10/1938	2,651.23	2,651.63	155.28
251	C. D. Tremor					
252	L. L. Slaughter					
254	L. D. Christian		02/10/1938	2,354.51	2,350.59	120.30
255	W. M. Johnson		02/10/1938	2,074.28	2,074.28	105.83
257	Kirk Walker	03/02/1937	01/26/1938	3,128.68	3,128.68	159.62
260	Coulter					
261	C. D. Tremor					
262	Dewey Cox					
263	Buster Dykes	04/17/1935	02/10/1938	2,067.21	2,067.21	105.49
264	John Bagby					
264	W. M. Carter		02/02/1938	2,184.13	2,184.13	111.43
265	A. W. Ledford	05/24/1935	01/12/1938	2,126.33	2,088.72	108.48
266	Ray Cash	03/24/1935	02/05/1938	2,183.60	2,142.06	111.41
273	Ben Hendrix	03/02/1936	01/12/1938	2,051.83	2,051.83	104.68
274	Jasper J. Allen	03/12/1936	02/10/1938	2,112.32	2,112.32	107.77
275	C. H. Creamer					
276	Ed Knight	02/18/1936	02/10/1938	2,225.36	2,181.92	113.54
277	Oscar Mitts		01/12/1938	2,045.72	2,045.72	104.37
278	M. M. Garrett					
279	Cecil Wilkins					
279	Newell Strawn		01/22/1938	2,202.80	2,202.80	112.39
280	Arnold Cox	03/30/1935	02/05/1938	2,094.87	2,055.03	106.88
281	Zeb Shaddix	03/26/1936	01/25/1938	2,337.18	2,325.01	119.24
283	Calvin Ramsey	03/27/1936	01/12/1938	2,033.35	2,033.35	103.74
288	Noah Passmore	02/27/1936	01/12/1938	2,097.80	2,097.80	107.03
289	Ben Johnson	02/27/1936	02/02/1938	2,003.20	1,981.72	102.20
289	Louie Sipes	03/15/1936				
290	Louie Sipes		08/05/1937	1,980.39	1,980.39	101.04
292	Walter L. Matlock	04/04/1936	08/05/1937	2,943.54	2,943.54	150.18
293	Quint J. Dent	03/31/1936	08/05/1937	2,995.71	2,995.71	152.84
294	Alford W. Shores					
295	Oscar L. Matlock	04/04/1936	08/05/1937	2,273.32	2,273.32	115.98
296	Bill J. King	03/31/1936	01/12/1938	2,573.78	2,573.78	131.31
297	Elmer Mooney	03/31/1936	02/22/1938	2,696.46	2,696.46	135.05
301	William Robert	03/28/1936	02/10/1938	2,111.30	2,111.30	112.82
302	Harvey McVay		02/05/1938	2,846.53	2,846.53	145.23
304	John S. Bell	05/05/1936	07/14/1937	2,034.84	2,034.84	103.82
306	Marvin J. Harmon	03/13/1936	02/03/1938	2,162.03	2,148.70	110.31
312	John W. Williams	03/10/1936	01/22/1938	2,016.67	2,016.67	102.89
314	Doss Haile	02/06/1936	08/05/1937	2,506.91	2,506.91	127.90
315	Dewey O. Fielder	03/12/1936	07/14/1937	2,792.30	2,792.30	142.46
316	Clyde E. N. Kern					
317	Herman C. Pierson	03/13/1936	02/10/1938	2,768.93	2,768.93	141.27
320	Monroe M. Hinesley	04/25/1936	08/05/1937	2,908.53	2,908.53	148.39
322	Edward R. Dodd	01/28/1936	08/05/1937	2,122.01	2,122.01	108.26
323	Lee McChesney	02/04/1936	01/12/1938	2,055.47	2,055.47	104.87
324	George Coulter	04/24/1936	02/02/1938	2,333.98	2,333.98	122.11

A New Deal in Dyess

House #	Purchaser	Date Arrived	Date of Sale	Original Contract	Balance	Annual Payment
325	Ben Buggar	02/27/1936	07/14/1937	2,471.51	2,471.51	126.09
326	Ernest T. Adair	04/08/1936	02/02/1938	2,170.03	2,170.03	110.71
327	O. E. Waldrip	04/10/1936	02/10/1938	2,587.27	2,587.27	132.00
328	Chester E. Pierce	04/10/1936	02/10/1938	3,059.38	3,059.38	156.09
329	H. E. Mulford		02/05/1938	2,691.79	2,691.79	137.33
335	Charlie Yarbrough					
362	W. L. Williams	01/03/1936	02/05/1938	2,949.56	2,949.56	150.48
363	Homer Chaney	04/16/1936	02/10/1938	2,239.20	2,239.20	114.24
364	C. L. Brooks					
365	Garnett L. Beggs	02/19/1936	01/22/1938	2,262.50	2,262.50	115.43
367	R. D. Lansdale		06/05/1937	3,026.55	3,02655	154.41
372	Pearson					
380	Hill					
380	James J. Hale	04/24/1936	02/10/1938	2,082.80	2,082.80	106.26
382	E. Clanton	01/13/1937	08/30/1938	2,142.86	2,142.86	109.33
383	W. H. Frits	04/10/1936	02/05/1938	2,113.92	2,113.92	107.85
384	Clinton Bowers	01/24/1936	08/05/1937	2,114.92	2,114.92	109.69
385	William H. Kindy	02/19/1936	08/05/1937	2,722.92	2,722.92	138.92
386	Clovis McDonald	01/04/1937	09/14/1938	2,138.74	2,138.74	109.12
386	Ernest Horton					
388	Erwin Bowls	01/28/1936	07/14/1937	2,717.16	2,717.16	138.63
389	William L. Jacobs	02/18/1936	11/04/1937	2,304.14	2,304.14	117.56
391	Charles F. Culwell	03/12/1936				
394	Charles F. Culwell		02/05/1938	2,537.63	2,537.63	129.47
395	Prentice Elmore					
397	Clarence East		02/10/1938	2,669.90	2,669.90	136.22
398	Park W. Tate		02/10/1938	2,761.28	2,761.28	140.88
399	D. C. Thrailkill		02/10/1938	2,853.73	2,853.73	145.60
400	J. F. Moseley					
405	K. H. Craig	04/15/1936	02/05/1938	3,072.62	3,072.62	156.76
406	Thomas M. Anderson	05/02/1936	07/14/1937	3,126.62	3,126.62	159.52
407	Richard K. Wooten	04/15/1936	12/18/1937	2,981.65	2,981.65	152.12
408	J. L. Jacobs	04/11/1936	12/18/1937	2,990.75	2,990.75	152.58
413	Fay O. Allen	05/02/1936	07/14/1937	2,940.36	2,940.36	150.02
414	Homer Johnson					
414	Etsel Hathcock		02/05/1938	2,770.27	2,770.27	141.34
415	A. G. Nichols	03/31/1936	02/02/1938	2,893.10	2,893.10	147.60
416	Dewey Smothers	04/19/1935	01/24/1938	2,709.88	2,709.88	138.26
417	W. T. Tapp					
421	Donald Dix					
424	Homer Johnson					
424	Fred Dallas	01/15/1937	07/26/1938	2,814.96	2,814.96	143.62
430	Jordan J. Smith	02/29/1936	02/05/1938	2,905.03	2,905.03	148.21
431	Wilson					
431	Harvey					
432	D. L. Wilson		06/07/1938	2,464.47	2,464.47	125.74
435	Everett J. Sutton	04/22/1936	02/10/1938	2,699.70	2,699.70	137.74

Appendix F

House #	Purchaser	Date Arrived	Date of Sale	Original Contract	Balance	Annual Payment
436	Charlie Yarbrough					
438	Matthew M. Creager	01/09/1937	07/26/1938	2,100.26	2,100.26	107.15
438	J. E. Southard					
439	Roy Ferguson					
440	Cyrus Fudge					
441	Robert A. Foley	03/13/1936	11/04/1937	2,551.59	2,551.59	130.18
442	Cyrus Fudge					
442	John Thompson	02/18/1936	02/10/1938	2,321.07	2,294.86	118.42
443	Roy Fergeson		02/05/1938	2,651.27	2,607.93	135.29
445	D. R. Alexandra					
446	Ira Looney					
448	G. W. Love	04/13/1937	07.26/1938	2,804.82	2,804.22	143.10
448	Tom Redding					
449	Loyd Ward		02/02/1938	2,675.53	2,675.53	136.50
451	Bud Wallace					
453	Ray Shepherd					
454	Alfred H. Campbell	04/01/1936				
455	Guy Nickols					
455	Clifford Pate		02/05/1938	2,891.43	2,891.43	147.52
458	Elmer Ivy					
459	James Cook		02/05/1938	2,691.59	2,691.59	137.32
460	Dewey Hudson		07/26/1938	2,793.01	2,793.01	142.50
462	Oilkington					
462	Cecil Mooney		07/26/1938	2,871.39	2,871.39	146.50
467	Homer Harlson		02/05/1938	2,733.44	2,676.45	139.46
468	Sam Massey					
469	J. W. Culpepper					
470	W. H. Smith		06/07/1938	2,639.73	2,639.73	134.68
516	K. D. Craig		02/10/1938	2,307.19	2,307.19	117.71
517	W. T. Tapp	02/06/1936	02/05/1938	2,036.14	2,036.14	103.88
520	Guy B. Williams	02/04/1936	02/05/1938	1,896.62	1,896.62	96.76
522	B. B. Wilson		02/05/1938	1,850.43	1,850.43	94.41
523	Buff Roberts		02/05/1938	1,905.01	1,905.01	97.19
535	Ed A. Ballentine	02/23/1936	07/14/1937	2,325.98	2,325.98	118.67
536	Charles Hill					
538	T. G. Mitchell		02/05/1938	2,692.79	2,692.79	137.38
539	J. H. Christmam					
540	B. F. Wright		02/05/1938	2,768.99	2,768.99	141.27
547	Sam Horton		02/05/1938	1,891.66	1,891.66	96.51
548	S. H. Wallace		02/05/1938	1,962.29	1,962.29	100.11
549	J. R. Clifton		02/05/1938	1,842.17	1,842.17	93.99
553	Earl Erwin					
561	Geo. Higgenbotham	03/28/1936	02/05/1938	2,496.92	2,496.92	127.39
562	Edger Pulliam					
564	Harry Myers		02/05/1938	2,000.57	2,000.57	102.07
565	Homer Clifton	03/26/1937	02/05/1938	1,884.29	1,884.29	96.14
567	D. C. Deaton					
567	Oscar Mabrey		10/17/1938	1,855.44	1,855.44	94.96

A New Deal in Dyess

House #	Purchaser	Date Arrived	Date of Sale	Original Contract	Balance	Annual Payment
568	Ernest A. Pickens	11/16/1935	02/05/1938	2,068.87	2,068.87	105.55
571	John M. McKaskle	03/18/1937	08/30/1938	2,023.68	2,023.68	103.25
574	William F. Henry	04/17/1936	08/05/1937	2,782.11	2,782.11	141.94
575	Bert L. Mathis	03/16/1936	06/05/1937	2,782.11	2,782.11	141.94
578	Don Smith	04/22/1936	02/05/1938	1,956.37	1,956.37	99.81
588	Frank Procter	04/29/1936	02/05/1938	2,810.70	2,810.70	143.40
590	Luther Warhurst	04/03/1937	07/26/1938	2,684.50	2,684.50	136.96
592	William Woods	03/16/1936	02/05/1938	2,514.81	2,514.81	128.30
594	B. J. Wilson	03/19/1936	02/05/1938	2,299.72	2,299.72	117.35
595	William F. Back	02/27/1936	02/05/1938	2,024.26	2,024.26	103.28
596	Alvin Taff	03/28/1936	01/12/1938	2,705.55	2,705.55	138.04
597	R. W. Daniel	03/02/1937	07/26/1938	2,078.64	2,078.64	106.05
598	U. S. Malot					
600	Alfred W. Shores	03/28/1936	02/05/1938	2,575.46	2,575.46	131.40
602	Homer C. McDaniel	03/12/1936	08/30/1938	2,641.24	2,641.24	134.75
607	R. N. Hovator	06/10/1935	01/12/1938	2,406.15	2,406.15	122.76
610	Bill Minor	04/16/1936				
611	Joe Wilson					
612	W. T. House					
614	S. H. Wallace	04/10/1936	02/10/1938	2,558.29	2,558.29	130.52
615	John Counts	04/14/1936	02/05/1938	2,363.51	2,363.51	120.58
617	D. M. Howard	03/09/1936	02/05/1938	2,567.69	2,567.69	131.00
618	Charles F. Wells	05/01/1936	01/12/1938	2,454.83	2.454.83	125.24
619	Richard A. Dotson	04/04/1936	02/10/1938	2,142.17	2,142.17	109.29
621	Harold Cline	03/03/1936	02/02/1938	1,748.34	1,748.34	89.20
622	Kinneb. Shoemaker					
625	James J. Darnell	03/28/1936	02/05/1938	2,045.53	2,045.53	104.36
626	Warren Wyscaver	04/16/1936				
627	Cecil B. Maddox	04/09/1936	12/18/1937	1,836.54	1,808.65	93.70
628	John T. Clore	04/24/1936	02/03/1938	2,560.12	2,560.12	130.62
629	Sam Bass					
630	Arthur Thorn					
630	Warren Wyscaver		02/05/1938	2,335.01	2,335.01	119.13
630	Norman James	04/16/1937				
632	Robert Hargett	04/17/1936				
633	Baxter Walden	04/08/1936				
634	Hershal O. Henson	03/31/1936				
635	Burton C. Wampler	04/22/1936				
636	Dale Simer	04/17/1936	02/05/1938	1,796.19	1,796.19	91.64
637	William M. Bell	03/28/1936	07/14/1937	1,935.28	1,935.28	98.74
				595,975.10	594,662.2	30,407.26

Source: National Archives, Record Group 96, Farmers Home Administration, AK-80.

APPENDIX G
Colonists Who Moved Away
As of March 9, 1937

NAME	DATE ARRIVED	DATE LEFT	ORIGIN COUNTY	REASON GIVEN FOR LEAVING
Goza, John	11/9/1934	3/12/1936	Arkansas	Man developed active tuberculosis
Billingsly, S. I.	1/30/1935	3/19/1936	St. Francis	Dissatisfied; objects to everything
Belk, Clarence	3/2/1936	4/29/1936	Miller	Wife and children not satisfied
Whitaker, S. I.	2/14/1936	5/6/1936	Faulkner	Dissatisfied--wife wants to stay
Long, Luther	3/28/1936	5/6/1936	Lee	Dissatisfied, health not good
Sams, J. D.	3/24/1935	5/27/1936	Miller	Illness of wife
Works, Bob H.	4/25/1936	5/29/1936	Cleveland	Sickness and dissatisfaction
Craig, A. R.	4/20/1935	5/29/1936	Jefferson	Death of wife, left w/2 small kids
Edwards, A. D.	3/16/1936	6/3/1936	Conway	Doesn't like colony
Atkins, R. B.	4/29/1936	6/10/1936	Ouachita	Domestic trouble
Rosenbaum, C. R.	3/2/1936	6/18/1936	Hempstead	Sickness in family
Lochals, W. M.	4/22/1936	6/24/1936	Ashley	Illness of father
Carlton, R. C.	3/26/1936	6/26/1936	Newton	Dissatisfied, thinks health not good
Smith, Clarence	3/30/1936	6/27/1936	Phillips	Wife ill and dissatisfied
Anderson, P. R.	4/17/1935	6/29/1936	Pulaski	Sick and discouraged
Cornellus, M. B.	3/26/1936	7/1/1936	Sevier	Back to mother-in-law's farm
Brantley, Terrell	2/27/1936	7/1/1936	Nevada	Told would get plenty of clothes, did not
Phillips, G. W.	2/13/1936	7/7/1936	Lincoln	Children sick
Brazier, R. W.	4/4/1936	7/9/1936	Yell	Doesn't have good health in colony
Boon, Elmer	4/30/1935	7/9/1936	Yell	Work offered on outside
Gibbs, Richard	1/31/1935	7/10/1936	Arkansas	Refused to sign mortgage
Plumlee, W. M.	2/6/1936	7/11/1936	Monroe	Sickness in family
Southworth, W. H.	3/20/1936	7/14/1936	Sevier	Illness of family
Cameron, A. J.	4/2/1936	7/14/1936	Union	Objects to mortgage & general plan
Litherland, Lee	3/7/1936	7/16/1936	Randolph	Wife developed active tuberculosis
Brooke, C. L.	1/28/1936	7/16/1936	Prairie	Thinks health not good on colony
Bryant, Milwee	4/7/1936	7/16/1936	Monroe	Does not like this country
Noe, Ray	4/1/1936	7/17/1936	Marion	Objects to mortgage
White, Luther	3/19/1936	7/17/1936	Grant	Objects to signing mortgage
Smith, Bryant	2/19/1936	7/17/1936	Grant	Dissatisfied and wife ill
Jones, Ed	4/22/1936	7/19/1936	Ashley	Dissatisfied after receiving bonus
Staton, Sebe	3/19/1936	7/20/1936	Arkansas	Objects to mortgage
Butler, Ivan	2/10/1936	7/20/1936	Mississippi	No complaints; wants to farm with uncle
Martin, Roy	3/16/1936	7/21/1936	Conway	Children health not good
Thomas, Freed	3/25/1936	4/21/1936	Faulkner	Water does not agree with family
Copeland, Hubert	3/6/1936	7/24/1936	Cross	To go back to railroad work
Bradley, Davis	3/19/1936	7/24/1936	Dallas	Believes health not good on colony
Penedist, G. M.	5/24/1935	7/31/1936	Drew	Refused to sign mortgage

A New Deal in Dyess

NAME	DATE ARRIVED	DATE LEFT	ORIGIN COUNTY	REASON GIVEN FOR LEAVING
Kraus, Norgus	5/24/1935	8/6/1936	Pope	Refused to sign mortgage
Handley, Bert	11/9/1935	8/10/1936	Desha	Objected to mortgage
Bearie, Tim	4/4/1936	8/12/1936	Yell	Says entire family having chills
Gore, Ruel	4/17/1936	1/15/1937	Sevier	Objects to cooperative
Gore, Alvin	4/17/1936	1/15/1937	Sevier	Objects to cooperative
Massey, Sam	2/19/1936	1/19/1937	Calhoun	Too much mud and water
Wilkington, Jess	4/17/1936	1/19/1937	Monroe	Family not satisfied
Crawford, I. I.	3/16/1936	1/19/1937	Conway	Family dissatisfied
Hanks, Sam	3/12/1935	1/19/1937	Cross	Not contented on colony
Best, Henry	1/28/1936	1/19/1937	Prairie	Offered work on outside
Smith, Earl	1/10/1937	1/19/1937	Mississippi	Left during high water, didn't ret.
Sohn, Opie	2/27/1936	1/20/1937	Van Buren	Wife dissatisfied
Little, Floyd	4/18/1936	2/12/1937	Van Buren	Objects to cooperative
Diley, W. T.	1/3/1936	2/20/1937	Mississippi	Better chance elsewhere
Freeman, James	3/4/1936	2/23/1937	Yell	Left without giving reason
Halford, B. R.	4/17/1935	2/23/1937	Clay	Dissatisfied
Reeves, Doxie	3/28/1936	2/24/1937	Lee	Can do better elsewhere
Johnson, Lantie	4/29/1936	2/26/1937	Calhoun	Objects to cooperative
Schmalhurst, Carl	3/15/1935	2/26/1937	Clark	High water, health of wife
Tillman, Max	3/20/1935	2/26/1937	Lawrence	Can do better elsewhere
Clay, Luther	3/16/1936	2/27/1937	Randolph	Wife sick and can do better
Wilkerson, R. M.	4/23/1936	2/27/1937	Pulaski	Objects to cooperative
Ashcraft, Mose	2/29/1936	2/27/1937	Grant	Sick and don't like mud
Beck, Robert	5/11/1935	2/27/1937	Logan	Objects to cooperative
Lovet, W. T.	3/23/1935	2/28/1937	Ashley	Afraid of high water
Hoskins, Herbert	5/19/1936	3/2/1937	Phillips	Could not conform to requirements
Deaton, Gilbert C.	12/12/1936	3/2/1937	Mississippi	To secure work outside farm
Dickinson, Tom	2/7/1936	3/2/1937	Mississippi	Don't like the place
Bennett, Frank	3/26/1936	3/3/1937	Van Buren	Objects to cooperative
Holland, Odie	3/26/1936	3/3/1937	Van Buren	Objects to cooperative
White, Allen C.	1/28/1936	3/3/1937	Craighead	Don't like price of farm/co-op
Bennett, C. E.	2/13/1936	3/3/1937	Lincoln	Dissatisfied
Wheatley, Mrs. Guy	2/5/1936	3/3/1937	Arkansas	Death of man
Satterfield, L. S.	3/26/1936	3/3/1937	Franklin	Man developed active tuberculosis
Harker, R. L.	5/31/1935	3/3/1937	Crawford	Death of man
Price, B. G.	3/30/1935	3/5/1937	Lawrence	To accept work in Texas
Moore, M. B.	4/6/1935	3/5/1937	Sevier	Dissatisfied; boy joined Navy
Mondy, M. P.	12/27/1935	3/6/1937	Mississippi	Could not conform to requirements
Rutherford, Frank	2/18/1936	3/7/1937	Grant	High water
Cook, Burl	4/3/1936	3/7/1937	Crawford	High water
Hines, Ed	11/25/1936	3/9/1937	Mississippi	Cannot comply with regulations
Haller, Veasey	4/22/1936	3/9/1937	Logan	High water
Lakey, Louis	3/4/1936	3/9/1937	Yell	Objects to cooperative

Source: National Archives, Record Group 96, Farmers Home Administration, AK-80.

Appendix H
Memories of a Lifetime
Participants Cited

CANNON JENNINGS, JEAN ANN (#2018-12-002)
Grandparents, Dale and Trudy Cunningham, were early settlers at Dyess from Bradley County, Arkansas. Her parents, Leroy and Elise (Cunningham) Cannon, became colony farmers in 1937-38. She was born at the Dyess hospital and graduated from high school in Dyess. Interviewers: Memories of a Lifetime Project Team, Arkansas State University, Oct. 20, 2017.

CASH YATES, JOANNE (#2018-11-001)
Sixth of the seven children of Ray and Carrie Cash. Discusses growing up in Dyess, memories of her brother Johnny Cash, and views of the project to restore the boyhood home. She was born in Dyess and graduated from the colony high school. Interviewers: Alex Brown's television production class, Arkansas State University, Aug. 4, 2011.

COX, LARRY (#2018-33-03)
Youngest child of Dewey and Elsie Cox, who moved to Dyess from Pike County in 1936 with seven children (Bernie, James, Lynn, Cohen, Joy and twins Janell and Lavell).. Their first child born at the Dyess Hospital died at three days old. Francine was born in June 1940, and Larry in June 1943 Interviewers: Ruth Hawkins, Heritage Sites, and Bryan Pierce, Heritage Studies Program, Arkansas State University, July 7, 2018.

FORRESTER WALLACE, FRANCES (#2018-10-002)
Born March 5, 1931 in Tyronza, moved to Dyess from Knobel with her parents, Victor and Clara Forrester, in 1945. She later taught in the Dyess Public Schools for 30 years and served as chairman of the Dyess reunion for more than 20 years. She died Sept. 19, 2009. Interviewer: Lisa Perry, Heritage Studies Program, Arkansas State University, May 19, 2007.

HENRY, WINFORD (#2018-12-005)
Son of William and Viola Henry who moved to Dyess in April 1936 from Faulkner County. His father, along with Ray Cash, helped rebuild many colony barns in the early 1940s after it was determined that original WPA-built barns were too small. Interviewers: Memories of a Lifetime Project Team, Arkansas State University, Oct. 20, 2017.

HENSON, A. J. (#2018-10-005)
Born in 1932, Henson moved to Dyess from Lawrence County in 1936 with his parents, Hershal O. and Nancy Kinder Henson, and siblings. He was in class with Johnny Cash and was a lifelong friend. A.J. left high school his junior year to enter military service. Interviewer: Lisa Perry, Heritage Studies Program, Arkansas State University, May 19, 2007.

HENSON, EVERETT (#2018-10-008)
Born Aug. 3, 1926, in Lawrence County, Henson moved to Dyess in March 1936 with his parents, Hershal O. and Nancy Kinder Henson. He had one older sister Freeda and five younger brothers, Howard, A. J., Coy Dean, Floyd, and Robert. He married Johnnie Murphy, also of Dyess. Henson was an unofficial historian for the town and maintained a website for former residents of Dyess. He died July 7, 2016, and left his Dyess materials to the Dyess Colony Archives. Interviewers: Elista Istre and Moriah Istre, Heritage Studies Program, Arkansas State University, May 12, 2010.

HOLLAND, ROBERT "BOB" (#2018-10-016)
Bob and his twin brother Bill were born in Siloam Springs, Arkansas, in 1924 to A. L. and Geneva Holland. He contracted infantile paralysis, and Franklin and Eleanor Roosevelt helped get him into the Warm Springs, Georgia, rehabilitation facility after his mother contacted them. When Dyess was developed Mrs. Roosevelt encouraged the Hollands to apply. They arrived in Dyess during 1936, just days before Mrs. Roosevelt's visit to the colony. Interviewer: Emmett Powers, Heritage Studies Program, Arkansas State University, June 14, 2013.

APPENDIX H

JOHNSON WRIGHT, HELEN (#2018-12-007)
Born in Etowah, Mississippi County, she moved to Dyess as a small child and graduated from Dyess High School in 1959. She played on the Dyess Eagles girls' basketball team, coached by Superintendent Lynn Cox. Interviewers: Memories of a Lifetime Project Team, Arkansas State University, Oct. 20, 2017.

KNIGHT CLEMENTS, VERA (#2018-10-014)
Daughter of Charlie T. and Agnes Knight who moved to Dyess in 1935 from Craighead County and remained until 1940. Her father was a farm laborer and a talented musician. He played in a band for a Mississippi County broadcast station. Interviewer: Emmett Powers, Heritage Studies Program, Arkansas State University, Feb. 20, 2013.

OWEN WISE, JANICE (#2018-10-018)
Born in south Arkansas and moved to Dyess Colony with her parents, Dewey and Elmer Annette Wise, during the 1930s. Her experience there was limited to three years between the family's arrival in 1934 and the 1937 flood. After the flood her family did not return except to pack up and move back to Columbia County, Arkansas. Interviewer: Emmett Powers, Heritage Studies Program, Arkansas State University, Sept. 18, 2013.

PHILLIPS, JAMES (#2018-33-004)
Son of Herschel and Lurly Phillips who moved from Dallas County in April 1936 when James was 9 years old. He left Dyess during his senior year in 1944 to enter the U. S. Navy in World War II. Interviewers: Ruth Hawkins, Heritage Sites, and Bryan Pierce, Heritage Studies Program, Arkansas State University, July 7, 2018.

A New Deal in Dyess

ROBERTS BLUE, ANN (#2018-12-010)
Arrived in April 1936 from Aubrey, Arkansas, with her parents Buff and Mattie Roberts and brother Marlin Joe. Two other brothers were born in Dyess: Herbert (born during the 1937 flood evacuation) and Carl Dwayne. The family moved to California in 1951. Interviewers: Memories of a Lifetime Project Team, Arkansas State University, Oct. 20, 2017.

WILSON MAULDIN, MARY LOU (#2018-12-011)
Daughter of Byron B. and Connie Wilson of Delight (Pike County), Arkansas. They lived in Dyess between 1934 and 1938 (from the time she was 6 until she was 10). Mauldin saved all her parents' receipts, contracts, and other papers from their time in Dyess and donated them to the Dyess Colony Archives. Interviewers: Memories of a Lifetime Project Team, Arkansas State University, Oct. 20, 2017.

WOOTEN, ED (#2018-10-001)
Born May 9, 1925, Ed was the son of Robert and Augusta Wooten, who moved to Dyess in 1936 from Lafayette County. He married Janie Barnes, also of Dyess. Wooten served as mayor of Dyess for 12 years and died Dec. 20, 2013. Interviewer: Lisa Perry, Heritage Studies Program, Arkansas State University, May 19, 2007.

WOOTEN, JANIE BARNES (#2018-10-001)
One of 12 children of John Sharp Barnes and Alva Etta Williams, who moved to Dyess around 1938 from the Bassett area. Her maternal grandparents Willie and Mittie Williams, were Dyess colonists who arrived in 1936 and also had 12 children. She married Ed Wooten and lives in Dyess. Interviewer: Lisa Perry, Heritage Studies Program, Arkansas State University, May 19, 2007.

Selected Bibliography

Books:

Agee, James. *Let Us Now Praise Famous Men*. Boston: Houghton Mifflin, 1941.

Albertson, Dean. *Roosevelt's Farmer: Claude R. Wichard in the New Deal*. New York: Columbia University Press, 1961.

Allen, Frederick Lewis. *Since Yesterday: The Nineteen-Thirties in America*. New York: Harper & Rowe, 1940.

Arkansas Emergency Relief Administration. *Traveling Recovery Road: The Story of Work-Relief and Rehabilitation in Arkansas, August 30, 1932 to November 15, 1936*. Little Rock, Ark., 1936.

Badger, Anthony J. *The New Deal: The Depression Years*. New York: Farrar Straus and Giroux, 1989.

Bagnall, Norma Hayes. *On Shaky Ground: The New Madrid Earthquakes of 1811-1812*. Columbia: University of Missouri Press, 1996.

Baldwin, Sidney. *Poverty and Politics: The Rise and Decline of Farm Security Administration*. Chapel Hill: University of North Carolina Press, 1968.

Barry, John M. *Rising Tide: The Great Mississippi Flood of 1927 and How It Changed America*. New York: Simon & Schuster, 1997.

Black, John D. *Agricultural Reform in the United States*. New York: McGraw-Hill, 1929.

Blackman, Douglas. *Slavery by another Name*. New York: Anchor Books, 2008.

Brunner, Edmund de S. and J. H. Kolb. *Rural Social Trends*. New York: McGraw-Hill, 1933.

A New Deal in Dyess

Caldwell, Erskine. *Tobacco Road*. New York: Signet, 1959.

Cantor, Louis. *Prologue to the Protest Movement: The Missouri Sharecropper Roadside Demonstration of 1939*. Durham, NC: Duke University Press, 1969.

Cash, Johnny, with Patrick Carr. *Cash: The Autobiography*. San Francisco: Harper Collins, 1997.

Cash, Wilbur J. *The Mind of the South*. New York: Knopf, 1941.

Cavin, James P., ed. *Economics for Agriculture*. Cambridge, Mass.: Harvard University Press, 1959.

Chandler, Lester V. *America's Greatest Depression, 1929-1941*. New York: Harper and Row, 1970.

Charles, Searle F. *Minister of Relief: Harry Hopkins and the Depression*. Syracuse, N.Y.: Syracuse University Press, 1963.

Conkin, Paul K. *FDR and the Origins of the Welfare State*. New York: Thomas Y. Crowell Company, 1967.

------. *Tomorrow a New World: The New Deal Community Program*. Ithaca, NY: Forgotten Books for Cornell University Press, 1959.

Conrad, David E. *The Forgotten Farmers: The Story of Sharecroppers in the New Deal*. Urbana: University of Illinois Press, 1965.

Dougan, Michael B. *Arkansas Odyssey: The Saga of Arkansas from Prehistoric Times to Present*. Little Rock, Ark.: Rose Publishing Company Inc., 1994.

Downs, William D., Jr. *Stories of Survival: Arkansas Farmers during the Great Depression*. Fayetteville, Ark.: Phoenix International, 2011.

Dykeman, Wilma and James Stokely. *Seeds of Southern Change: The Life of Will Alexander*. Chicago: University of Chicago Press, 1962.

Eaton, Joseph W. and Saul M. Katz. *Research Guide on Cooperative Group Farming*. New York: H. W. Wilson Company, 1942.

Edrington, Mabel F., ed. *History of Mississippi County, Arkansas*. Ocala, Fla.: Ocala Star Banner, 1962.

Eliot, T. S. "Dry Salvages." *Four Quartets*. New York: Harcourt Brace, 1941.

Flint, Timothy. *A Condensed Geography and History of the Western States or the Mississippi Valley, Volume 1*. Gainesville, Fla.: Scholars' Facsimiles and Reprints, 1970.

Freidel, Frank. *Launching the New Deal*. Boston: Little, Brown, 1973.

Galbraith, John Kenneth. *The Great Crash: 1929*. New York: Houghton Mifflin Company, 1997.

Genung, A. B. "Agriculture in the World War Period." *Farmers in a Changing World: The Yearbook of Agriculture, 1940*. Washington, D.C., 1940.

Gillin, John L. *Poverty and Dependency*. New York: Century, 1926.

Goodwin, Doris Kearns. *Franklin and Eleanor Roosevelt: The Home Front in World War II*. New York: Touchstone, 1994.

Govoni, Albert. *A Boy Named Cash*. New York: Lancer Books, 1970.

Graustein, Jeannette. *Thomas Nutall Naturalist: Explorations in America 1808-1841*. Cambridge, Mass.: Harvard University Press, 1967.

Gregory, James N. *The Dust Bowl Migration and Okie Culture in California*. New York: Oxford University Press, 1989.

Grubbs, Donald. *Cry From the Cotton: The Southern Tenant Farmers' Union and the New Deal*. Chapel Hill: University of North Carolina Press, 1971.

Hawkins, Van. *Plowing New Ground: The Southern Tenant Farmers Union and Its Place in Delta History*. Virginia Beach, Va.: Donning, 2007.

Hibbard, Benjamin H. *Effects of the Great War Upon Agriculture in the United States and Great Britain*. New York: Oxford University Press, 1919.

Holley, Donald. *Uncle Sam's Farmers: The New Deal Communities in the Lower Mississippi Valley*. Urbana: University of Illinois Press, 1975.

Hoover, Herbert. *The Memoirs of Herbert Hoover: The Great Depression, 1929-1949*. New York: Macmillan, 1952.

Hopkins, Harry L. *Spending to Save: The Complete Story of Relief*. New York: Norton, 1936.

Infield, Henrik F. *Cooperative Communities at Work*. New York: Dryden Press, 1945.

A New Deal in Dyess

Johnson, Charles Spurgeon, Edwin R. Embree, and Will Winton Alexander. *The Collapse of Cotton Tenancy*. Chapel Hill: University of North Carolina Press, 1935.

Kazin, Alfred. *On Native Grounds*. Garden City, N Y: Doubleday, 1942.

Kennedy, David M. *The American People in the Great Depression: Freedom from Fear, Part One*. New York: Oxford University Press, 1999.

Kester, Howard. *Revolt Among the Sharecroppers*. New York: Arno Press, 1969.

Key, V.O. *Southern Politics in State and Nation*. New York: Knopf, 1949.

Kirby, Jack Temple. *Rural Worlds Lost: The American South, 1920-1960*. Baton Rouge: Louisiana State University Press, 1987.

Kirkendall, Richard S. *Social Scientists and Farm Politics in the Age of Roosevelt*. Columbia: University of Missouri Press, 1966.

Loomis, Charles P. and J. Allan Beegle. *Rural Social Systems*. New York: Prentice-Hall, 1950.

Lord, Russell. *The Agrarian Revival*. New York: American Association for Adult Education, 1939.

Lowittj, Richard and Maurine Beasley, eds. *One Third of a Nation: Lorena Hickok Reports on the Great Depression*. Urbana: University of Illinois Press, 1981.

Magnum, Charles S. *The Legal Status of the Tenant Farmer in the Southeast*. Chapel Hill: University of North Carolina Press, 1952.

Maris, Paul V. *The Land Is Mine*. Washington, D.C.: GPO, 1950.

Miller, Marc S. ed. *Working Lives: the Southern Exposure History of Labor in the South*. New York: Pantheon, 1980.

Mitchell, Broadus. *Depression Decade: From New Era through New Deal*. Armonk, NY: M. E. Sharpe, 1975.

Mitchell, H. L. *Mean Things Happening in This Land: The Life and Times of H. L. Mitchell, Co-Founder of the Southern Tenant Farmers Union*. Montclair, N.J.: Allanheld, Osmun & Co., 1979.

Murchison, Claudius T. *King Cotton is Sick*. Chapel Hill: University of North Carolina Press, 1930.

Myers, William Starr and Walter H. Newton. *The Hoover Administration: A Documented Narrative*. New York: Charles Scribner's Sons, 1936.

Nixon, H. C. *Forty Acres and Steel Mules*. Chapel Hill: University of North Carolina Press, 1938.

Pells, Richard H. *Radical Visions, American Dreams*. New York: Harper and Row, 1973.

Rasmussen, Wayne D. *Agriculture in the United States: A Documentary History*. New York: Random House, 1975.

Rochester, Anna. *Why Farmers Are Poor*. New York: International Publishers, 1940.

Saloutos, Theodore. *The American Farmer and the New Deal*. Ames: Iowa State University Press, 1982.

Schlesinger, Arthur M., Jr. *The Age of Roosevelt: The Crisis of the Old Order, 1919-1933*. Boston: Houghton Mifflin, 1957.

------. *The Coming of the New Deal*. Boston: Houghton Mifflin, 1959.

------. *The Politics of Upheaval*. Boston: Houghton Mifflin, 1960.

Sherwood, Robert Emmet. *Hopkins and Roosevelt: An Intimate History*. New York: Harper and Brothers, 1950.

Shlaes, Amity. *The Forgotten Man: A New History of the Great Depression*. New York: Harper Collins, 2007.

Steinbeck, John. *The Grapes of Wrath*. New York: Viking, 1939.

Sternsher, Bernard. *Rexford Tugwell and the New Deal*. Brunswick, N.J.: Rutgers University Press, 1964.

Streissguth, Michael. *Johnny Cash: The Biography*. Cambridge, Mass.: Da Capo Press, 2006.

------. *Ring of Fire*. Cambridge, Mass.: Da Capo Press, 2003.

Tindall, George B. *The Emergence of the New South, 1913-1945*. Baton Rouge: Louisiana State University Press, 1967.

Tugwell, Rexford G. *The Battle for Democracy*. New York: Columbia University Press, 1935.

------. *The Brains Trust*. New York: Viking, 1968.

------. *The Democratic Roosevelt*. Garden City, N.Y.: Doubleday, 1957.

Turner, Steve. *The Man Called Cash: The Life, Love, and Faith of an American Legend*. Nashville, Tenn.: Thomas Nelson, Inc., 2005.

Wallace, Henry A. *New Frontiers*. New York: Reynal & Hitchcock, 1934.

Watkins, T. H. *The Great Depression: America in the 1930s*. New York: Backside, Inc., 1993.

Whayne, Jeannie M. *Delta Empire: Lee Wilson and the Transformation of Agriculture in the New South*. Baton Rouge: Louisiana State University Press, 2011.

------. *A New Plantation South: Land, Labor, and Federal Favor in Twentieth Century Arkansas*. Charlottesville: University of Virginia Press, 1996.

------, and Willard B. Gatewood. *The Arkansas Delta: Land of Paradox*. Fayetteville: University of Arkansas Press, 1993.

Williams, C. Fred. *Arkansas Independent and Proud*. Sun Valley, Calif.: American Historical Press, 2002.

Wilson, Edmund. *The American Earthquake: Documentary of the Twenties and Thirties*. Garden City, N.Y.: Doubleday, 1958.

Woodruff, Nan Elizabeth. *American Congo: The African American Freedom Struggle in the Delta*. Cambridge, Mass.: Harvard University Press, 2003.

Workers of the Writers' Program of the Work Projects Administration in the State of Arkansas. *Arkansas: A Guide to the State*, 1941. New introduction by Elliott West. Lawrence: University Press of Kansas, 1987.

Wren, Christopher S. *Winners Got Scars Too: The Life of Johnny Cash*. New York: Ballantine Books, 1971.

JOURNALS AND PERIODICALS:

"Arkansas Experiment." *Commonweal*, 21 (November 16, 1934), 95.

"Arkansas's Fight for Life." *Literary Digest*, 108, 6.

Blackwell, Gordon W. "The Displaced Tenant Farm Family in North Carolina." *Social Forces*, 13 (October 1934), 65-73.

BIBLIOGRAHY

Clifton, Maxine Mitchell. "Thomas Green," *Delta Historical Review*. Mississippi County Historical and Genealogical Society, Mississippi County, Arkansas, 2:1 (Summer 1990), 20-21.

Colcord, Joana C. "Training for Intentional Community." *Alternate Society* (July 1970), 8-9.

"Good News from Arkansas." *Literary Digest*, 108, 12.

Hamilton, Don. "Jonesboro and Arkansas during the Flood of 1937." *Craighead County Historical Quarterly*, 9 (Winter 1971), 2-11.

Hammar, C. H. "An Appraisal of Resettlement: Discussion." *Journal of Farm Economics*, 19:1 (February 1937), 202-205.

Harris, Lement. "An Arkansas Farmer Speaks." *New Republic*, 67 (May 27, 1931), 40-41.

Hayden, David. "A History of Dyess, Arkansas." M. A. Thesis, Department of History, Southern Illinois University. August 1970.

Henson, Everett Dewey. "Memories of Dyess Colony." *Delta Historical Review*. Mississippi County Historical and Genealogical Society, Mississippi County, Arkansas, 2:1 (Summer 1990), 3-21.

Hicks, Floyd W. and C. Roger Lambert. "Food for the Hungry: Federal Food Programs in Arkansas, 1933-1942." *Arkansas Historical Quarterly*, 37 (Spring 1978), 23-43.

Holley, Donald. "Trouble in Paradise: Dyess Colony and Arkansas Politics." *Arkansas Historical Quarterly*, 32 (Autumn 1973), 203-216.

Hopkins, Harry. Hopkins press highlights, Sept. 27, 1935. n.p., WPA press clippings, Box 5, Arkansas State Archives.

Hudgens, R. W. "The Plantation South Tries a New Way." *Land Policy Review*, 3 (November 1949), 26-29.

Johnson, Charles S. "Incidence of the New Deal Programs upon the Negroes." *American Journal of Sociology*, 11 (May 1935), 737-745.

Johnson, Homer Joe. *Delta Historical Review*. Mississippi County Historical and Genealogical Society, Mississippi County, Arkansas. 2:1 (Summer 1990), 15-17.

Kirkpatrick, E. L. "Housing Aspects of Resettlement." *Annals of the American Academy of Political and Social Science*, 190, Current Developments in Housing (March 1937), 94-100.

Koch, Lucien. "War in Arkansas." *New Republic*, 82 (March 27, 1935), 183.

Lambert, Roger. "Hoover and the Red Cross in the Arkansas Drought of 1930." *Arkansas Historical Quarterly*, 29 (Winter 1970), 3-19.

Lohof, Bruce A., ed. "Herbert Hoover's Mississippi Valley Land Reform Memorandum: A Document." *Arkansas Historical Quarterly*, 29 (Spring 1970), 112-119.

Loomis, Charles P. and Dwight M. Davidson, Jr. "Social Agencies in the Planned Communities." *Sociometry*, 2:33 (July 1939), 24-42.

Mead, Elwood. "Community Farming." *New Republic*, 41 (1924-1925), 329-330.

Mitchell, H. L. "The Founding and Early History of the Southern Tenant Farmers Union." *Arkansas Historical Quarterly*, 32 (Winter 1973), 342-369.

Murray, Gail S. "Forty Years Ago: The Great Depression Comes to Arkansas." *Arkansas Historical Quarterly*, 29 (Winter 1970), 293-310.

Osborn, George C. "The Southern Agricultural Press and Some Significant Rural Problems." *Agricultural History*, 29 (July 1955), 115-122.

Pittman, Dan W. "The Founding of Dyess Colony." *Arkansas Historical Quarterly*, 29 (Winter 1970), 313-325.

Randall, Mark. "The Best Proposition A Poor Man Ever Had: The Founding of Dyess Colony." Wilkerson-Freeman (Spring 2010), Hist 6282, U. S. in Crisis 1929-1949.

Rison, David. "Federal Aid to Arkansas Education, 1933-1936." *Arkansas Historical Quarterly*, 36 (Summer 1977), 192-200.

Simonson, S. E. "The St. Francis Levee and High Waters on the Mississippi River." *Arkansas Historical Quarterly*, 6 (Winter 1947), 417-421.

------. "Origin of Drainage Projects in Mississippi County." *Arkansas Historical Quarterly*, 5 (Autumn 1946), 263-265.

Simpson, Roy Vergil. "Reminiscences of a Hill Country School Teacher." *Arkansas Historical Quarterly*, 27 (Summer 1968), 159.

Tugwell, Rexford G. "Farm Relief and a Permanent Agriculture." *The Annals of the American Academy of Political and Social Science*, 142 (March 1929), 271-282.

------. "The Resettlement Idea." *Agricultural History*, 33 (October 1959), 159-164.

Vance, Robert B. "How the Other Half is Housed." *Southern Policy Papers*, 4, University of North Carolina Press, 1936.

Vaughn, Burton F. "Arkansas Makes a Brilliant Recovery." *Review of Reviews*, 83 (June 1931), 90.

Venkataramani, M. S. "Norman Thomas, Arkansas Sharecroppers, and the Roosevelt Agricultural Policies, 1933-1937." *Arkansas Historical Quarterly*, 24 (Spring 1965), 3-28.

Wade, Mike. "Founding of Dyess Colony." August 1979 manuscript chapter in possession of Jean Ann Cannon/Jennings. Posted on Dyess Colony private website, June 21, 2005.

Webb, Pamela. "By the Sweat of the Brow: The Back-to-the-Land Movement in Depression Arkansas." *Arkansas Historical Quarterly* 42: 4 (Winter 1983), 332-345.

Wherry, Elizabeth C. "A Chance for the Share-Cropper: Farm Colony at Dyess, Arkansas Provides a New Start for Farmers Beaten By Depression." *Wallaces' Farmer and Iowa Homestead*, 63 (May 7, 1988), 5.

Wilson, Charles Morrow. "Famine in Arkansas." *Outlook*, 57 (April 29, 1931), 596.

PERSONAL HISTORIES:

Blake, Henry. Interview with Samuel S. Taylor. "Born in Slavery: Slave Narratives from the Federal Writers Project, 1936-1938." Arkansas Narratives, 2:1, Manuscript Division, Library of Congress.

Burkhart, Bernice Alma. Personal unpublished history. Dyess Colony Archives #2018-13-001, Arkansas State University.

Cash, Johnny. Excerpt from interview conducted by Ed Salamon, program director, WHN Radio, New York. *The Johnny Cash Silver Anniversary Radio Special*, aired on the Mutual Broadcasting System. July 4, 1980.

Cox, Larry. "Stories from Dyess." Posted October 27, 2004, on Dyess Colony private website, administered by A. J. and Everett Henson. Hard copy in Everett Henson Collection, Dyess Colony Archives, Arkansas State University.

Gordey, Malcomb. Letter to Everett Henson, July 6, 2000. Posted on Dyess Colony private website, A. J. and Everett Henson, admin. Hard copy in Everett Henson Collection, Dyess Colony Archives, Arkansas State University.

Hardin, Ed. "Fun Things to do in the Early Years of Mississippi County." Compiled by Jean Ann Cannon Jennings. Posted on Dyess Colony private website, administered by A. J. and Everett Henson. Hard copy in Everett Henson Collection, Dyess Colony Archives, Arkansas State University.

Roberts, Marlin Joe. Personal unpublished history prepared for Dyess reunion, July 11, 2005. Dyess Colony Archives #2018-13-002, Arkansas State University.

Story of the Mississippi County Project. As Told by a Man on the Job, June 1934, National Archives Record Group 69.

PUBLIC DOCUMENTS:

Arkansas Department of Social Services. "Report on Colonization Project No. 1." April 17, 1935. Everett Henson Collection, Dyess Colony Archives, Arkansas State University.

"Articles of Incorporation, Dyess Colony Cooperative Association." October 1936. Everett Henson Collection, Dyess Colony Archives, Arkansas State University.

Background Section, "Final Report and Physical Accomplishments of the Arkansas Work Projects Administration." March 1, 1943. Everett Henson Collection, Dyess Colony Archives, Arkansas State University.

Brown, Russell and Company. "Report to the Office and Directors, Dyess Colony, Incorporated." Little Rock, Ark., February 29, 1936. Everett Henson Collection, Dyess Colony Archives, Arkansas State University.

Dudley, E. S., Letter to Col. Lawrence Westbrook. March 19, 1937. Record Group 96, Farmers Home Administration, National Archives.

Dudley, E. S., Memorandum to Dyess colonists. December 17, 1936. Everett Henson Collection, Dyess Colony Archives, Arkansas State University.

"Dyess Colony, Inc. Project Book," Chapter III. Community Operation and Management, Section F. Education, Recreation and Health, Religious Groups.

BIBLIOGRAHY

Records Management Division, Resettlement Administration. Record Group 96, Farmers Home Administration, AK-80, 700, National Archives.

Finch, Paul. Letter to the editor, *Memphis Press-Scimitar*, April 20, 1938. Work Projects Administration, Special Collections, University of Arkansas Libraries, Box 10, File 87.

Gallagher, Eula. "Resume of Procedure and Problems of Planning Inspection Trips and Moving Families to Dyess Colony." Family Selection Section, Resettlement Administration, April 25, 1936. Record Group 96, Farmers Home Administration, AK-8, National Archives.

Holt, John P. "An Analysis of Methods and Criteria used in Selecting Families for Colonization Projects." Social Research Report No. 1, Farm Security Administration and the Bureau of Agricultural Economics, Washington, D.C., 1937. Record Group 96, Farmers Home Administration, National Archives.

Hopkins, Harry. Letter to Floyd Sharp, August 24, 1936. Record Group 96, Farmers Home Administration, National Archives.

Johnson, Mary E. "Preliminary Report on Community Development Policy." July 3, 1936. Record Group 96, Farmers Home Administration, National Archives.

McChesney, Lee and Arkansas Rural Rehabilitation Corporation. Contract. April 16, 1935. Records of the Dyess Colony, Farmers Home Administration, Osceola, Arkansas. Dyess Colony Archives, Arkansas State University.

Maris, Paul V. "Rehabilitation Policies." Summary by Supervisor of Field Services, Arkansas Federal Emergency Relief Administration, W355-21. Everett Henson Collection, Dyess Colony Archives, Arkansas State University.

Sharp, Floyd. Memorandum to All Residents of the Dyess Colony. February 10, 1936. Works Progress Administration, Little Rock, Ark., WPA-674. In Dyess Colony, Inc. Project Book, Exhibit E, Chapter III, Section G. Terms of Occupancy. Dyess Colony Archives, Arkansas State University.

------. Memorandum to All Colony Families. December 30, 1938. Everett Henson Collection, Dyess Colony Archives, Arkansas State University.

------. Letter to Col. Lawrence Westbrook, April 29, 1938. Work Projects Administration, Special Collections, University of Arkansas Libraries, Box 10.

------. Letter to Senator Hattie Caraway, June 6, 1938. Work Projects Administration, Special Collections, University of Arkansas Libraries, Box 10, File 87.

------. Letter to WPA Office, Washington D.C., April 24, 1939. Work Projects Administration, Special Collections, University of Arkansas Libraries, Box 10, File 88.

Tapping, Amy Pryer. "Report on Colonization Project No. 1." Emergency Relief Administration Social Service Division, March 25, 1935. Everett Henson Collection, Dyess Colony Archives, Arkansas State University.

U. S. Congress. A Bill to Help Tenant Farmers Become Land-Owners. H.R. 7562, 75th Congress, 1st Session. July 13, 1937.

U. S. Farmers Home Administration. "Records of Dyess, Arkansas, 1934-1952." Washington, D.C.: National Archives, Record Group 96, Boxes 50-67.

U. S. Works Progress Administration. "Records of Dyess, Arkansas, 1934-1940." Washington, D.C.: National Archives, Microfilm, three rolls, 1744-1746.

Westbrook, Lawrence. "Rural Industrial Communities for Stranded Families." Federal Emergency Relief Administration, 1934.

INDEX

1811-12 earthquakes, 15
1927 flood, 20, 23-24
1929 Stock Market Crash, 25, 28, 35
1930-31 drought, 27-29, 35
1936 Anniversary Celebration, 87, 89
1937 flood, 90-95
Agricultural Adjustment Act (AAA), 23, 31-32, 41
Agricultural Defense Relations, 118
Agricultural Wheel, 36
American Farm Bureau Federation, 71
Arkansas Department of Social Services, 47, 49
Arkansas Relief Administration, 41
Arkansas Rural Rehabilitation Corporation (ARRC), 40-41, 86
Arkansas State University, 12
Arkansas Welfare Department, 49
Bailey, Arkansas Gov. Carl E., 97, 107
Bailey, Willie, 126-127
Bankhead-Jones Farm Tenancy Act, 117
"Big River Blues," 92
Blake, Henry, 17-18
Blue, Ann Roberts, 43, 66, 73, 120
Bullard, Velma, 104
Burkhart, Bernice Alma (Mrs. Allie), 63
   Allie, 63
   Travis, 63
Butler, Ben 44
Camp Pike, 91
Caraway, Sen. Hattie, 104
Carter family, 52
Cash family, 9, 13, 51, 93, 116, 120
   Carrie (Mrs. Ray), 13, 93, 116, 121
   Jack 120-121
   Joanne, 12, 51-52, 116
   Louise, 116

   Ray, 41, 51, 92, 103, 116, 120-121
   Reba, 116
   Roy, 74, 92, 116
   Tommy, 116
Cash, Johnny, 9, 13, 44, 51, 66, 74, 92-93, 120-121, 124-125
Civil War, 17, 24, 36
Civil Works Administration (CWA), 38
Civilian Conservation Corps (CCC), 37, 118
Clements, Vera Knight, 59
Cleveland County, Arkansas, 13
Clifton, Maxine Mitchell, 56
Colonization Project No. One, 11, 40
Colony Herald, 74, 78, 84, 88, 121, 126
Colvin, Milton, 87
Commodity Credit Corporation, 37, 119
Cotton Convention of 1926, 23
Craig, Arthur, 74
Creamery Package Company, 40
Crowley's Ridge, 16
Current River, 58
Currie, F. Arthur, 77
Daniels, Jonathan, 44, 77
Deaton family, 88
DeValls Bluff, Arkansas, 55
Division of Rural Rehabilitation (RR), 30
Division of Subsistence Homesteads, 30
Dyess, William R., 11, 40, 73, 123
   Billy, 73
Drew County, Arkansas, 55
Dudley, E. S., 78, 83, 87, 90-92, 96, 102-103
Dyess Colony Cooperative Association (DCCA), 87, 104-106, 137
Dyess Colony Incorporated (DCI), 71, 83, 97, 104-106, 108
Dyess Cooperative Gin Association, 117

A New Deal in Dyess

Dyess Cooperative Store Association, 117
Dyess Farms Incorporated (DFI), 117, 120
Dyess Medical and Health Association, 117
Dyess Rural Rehabilitation Corporation (DRRC), 97, 117, 120
Echols family, 78
Edmonston, Clarence, 126-127
Eichenbaum, Howard, 43-44, 50, 74
Emergency Relief Administration, 30, 48, 70
Eudy, Georgia Beatrice, 118
Fairview Plantation, 36
Farm Credit Administration (FCA), 37
Farm Security Administration (FSA), 30
Farm Service Agency (FSA), 30, 117, 120
Farmers and Exchange and Loan Company, 105
Federal Drought Relief Act, 29
Federal Emergency Relief Administration (FERA), 30-32, 37, 40-41, 44, 47-49
Finch, Paul, 103-104
"Five Feet High and Rising," 92
Flaherty, Anna, 119
Food For Victory, 118
Forrest City, Arkansas, 55
Funk, S. B., 107-109
Futrell, Gov. J. M., 41, 73, 97, 107
Future Farmers of America (FFA), 114
Gardner, Jim, 74
Geater, Elbert, 149
Gordey, Malcolm, 113
Grange, The, 36
Gray, Dan T., 41, 72
Hardin, Ed, 64
Hawkins, Ruth, 12
Hays, Arkansas Congressman Brooks, 32
Helena, Arkansas, 16
Henry, Winford, 114
Henson, A. J., 50-51, 111
Henson, Everett, 12, 111-112, 123
Henson, Hershal, 60, 126-127
   A. J. and Coy, 60
Hickok, Lorena, 37
Hinesley, Rosnel, 74
Historic Dyess Colony: Johnny Cash Boyhood Home, 9

Hitt, Arkansas, 42, 70
Holland, Robert (Bob), 79-81
   Alva, 79-81
   Geneva (Mrs. Alva), 79-81
Hollingsworth, Dr. 114
Home Improvement Association, 98
Hoover, President Herbert, 21, 25, 28-29
Hopkins, Harry 31, 37, 40, 83, 97
Hurst, Richard, 126-127
Jennings, Jean Ann Cannon, 65, 115
Johnson, Homer Joe, 55-56, 61-62, 64
Johnson, Mary E., 102
Keiser, Arkansas, 113
Kester, Howard, 32
Kimbrough, Hazel, 60
Kinsley, Phil, 77
Knobel, Arkansas, 62
Lee, Willie, 119
Lee Wilson and Company, 40
Lee Wilson Plantation, 112
Little Rock, Arkansas, 28, 42-43, 91, 108
Live Wire Club, 75
Lower Mississippi Valley, 24
Luxora, Arkansas, 41
Marie, Arkansas, 112
Marked Tree, Arkansas, 36
Mauldin, Mary Lou Wilson, 43, 115
McCravin, A. J., 107
McGuire, Superintendent, 113
Memories of a Lifetime, 8-9
Miller, John E., 97
Mississippi County, Arkansas, 6, 11, 18, 29, 40-42, 56, 64, 77, 107, 118
Mississippi River, 11, 15, 23-24, 40, 90, 92
Mitchell, H. L., 36-37
Mitchell, Thomas Green, 126-127
   Josephine and Maxine, 54
   Mary (Mrs. Thomas G.), 54
Murphy, Cone, 42, 158
National Better Homes Week, 78
National Defense Program, 118
National Register of Historic Places, 123
Native Americans, 15
Nimmons, Arkansas, 35
Norment, O. G., 41
Nuttall, Thomas, 15
Old Center, 43, 55, 91
Osceola, Arkansas, 44, 106

INDEX

Owen, Dewey, 126-127
Phillips family, Roscoe and Pearl, 122
   Dorothy Virginia, 122
   James Roscoe, 122
Phillips, James, 114
Pine Bluff, Arkansas, 83, 93
Presbyterian Board of Home Missions, 65
RAMA 21 form, 58
Red Cross, 29, 96
Reid, T. Ray, 41
Resettlement Administration (RA), 30, 33, 49, 57, 65, 71-72, 102
Robert H. McNair Jr, 41
Roberts family
   Ann, 43, 66, 73, 120, 149
   Dewayne, 149
   Herbert, 149
   Marlin Joe, 56, 149
Roosevelt, Eleanor, 76, 78-80
Roosevelt, President Franklin D., 9, 23, 29-32, 71, 79-80, 106
Russell Brown & Company, 70-71, 108
Salyers, Fern, 67, 78, 158
Sebastian County, Arkansas, 72
Sharp, Floyd, 41, 71, 74, 78, 83, 97, 102-104, 107
Shawnee, Arkansas, 113
Siloam Springs, Arkansas, 79, 80
Smith, Burl, 114
Smith, Oscar, 74
Smith, William Harve, 123
Southern Tenant Farmers Union (STFU), 34, 36-37, 52
Spicer, H., 67
St. Francis River, 55
Stone County, Arkansas, 56
Tapping, Amy Pryer, 48
Taylor, Dr. Carl, 33
Terry, Jake, 102
Truman, President Harry, 120
Tugwell, Rexford, 30, 49, 83
Turner, Lulu, 72
Tyronza River, 40, 42, 64, 66, 90, 91
U. S. Congress, 120
U. S. Department of the Interior, 24, 30
Wallace, Frances Forrester, 62-63
Wallace, Henry A., 31-32

Westbrook, Lawrence, 40, 49, 74, 83-87, 90, 92, 96, 102-103
Williams, Clyde, 119
Williams, David R., 44
Wilson, Arkansas, 113
Wilson, Byron B., 126-127
Wilson, Lee Jr., 44
Wilson, Mr. and Mrs. Lafayette, 56
Wilson, President Woodrow, 22
Women's Army Auxiliary Corps (WAAC), 119
Wood, M. A., 41
Wooten, Ed, 58
Wooten, Homer, 75
Works Progress Administration (WPA), 30, 37, 40, 49, 54, 71, 83-84, 97, 102, 107
World War I, 18, 21-22, 25, 93
World War II, 11, 118-120
Wright, Helen Johnson, 64
Yancey, Emmett, 126-127

www.ingramcontent.com/pod-product-compliance
Lightning Source LLC
Chambersburg PA
CBHW020652300426
44112CB00007B/347